CAMBRIDGE LIBRARY COLLECTION

Books of enduring scholarly value

Technology

The focus of this series is engineering, broadly construed. It covers technological innovation from a range of periods and cultures, but centres on the
technological achievements of the industrial era in the West, particularly
in the nineteenth century, as understood by their contemporaries. Infra-
structure is one major focus, covering the building of railways and canals,
bridges and tunnels, land drainage, the laying of submarine cables, and the
construction of docks and lighthouses. Other key topics include developments in industrial and manufacturing fields such as mining technology,
the production of iron and steel, the use of steam power, and chemical
processes such as photography and textile dyes.

The Life of Robert Stephenson, F.R.S.

Relying on incremental experiment and practice rather than individual leaps
into the unknown, Robert Stephenson (1803–59) forged an influential career
as a highly respected railway and civil engineer. From the steam locomotive
Rocket to the London and Birmingham Railway and the Britannia Bridge,
his work helped to consolidate the foundations of the modern engineering
profession. Based on the first-hand testimony of relatives and contemporaries
as well as correspondence and official records, this 1864 biography by John
Cordy Jeaffreson (1831–1901), published only five years after Stephenson's
death, tells the story of this quiet industrial innovator. Five chapters by
engineer William Pole (1814–1900) provide a more technical insight,
examining some of Stephenson's most significant railway bridges. Volume 2
covers his advocacy of standardisation of the permanent way during the
Gauge War, and his life as a bridge builder and politician.

Cambridge University Press has long been a pioneer in the reissuing of out-of-print titles from its own backlist, producing digital reprints of books that are still sought after by scholars and students but could not be reprinted economically using traditional technology. The Cambridge Library Collection extends this activity to a wider range of books which are still of importance to researchers and professionals, either for the source material they contain, or as landmarks in the history of their academic discipline.

Drawing from the world-renowned collections in the Cambridge University Library and other partner libraries, and guided by the advice of experts in each subject area, Cambridge University Press is using state-of-the-art scanning machines in its own Printing House to capture the content of each book selected for inclusion. The files are processed to give a consistently clear, crisp image, and the books finished to the high quality standard for which the Press is recognised around the world. The latest print-on-demand technology ensures that the books will remain available indefinitely, and that orders for single or multiple copies can quickly be supplied.

The Cambridge Library Collection brings back to life books of enduring scholarly value (including out-of-copyright works originally issued by other publishers) across a wide range of disciplines in the humanities and social sciences and in science and technology.

The Life of
Robert Stephenson, F.R.S.

*With Descriptive Chapters on Some
of his Most Important Professional Works
by William Pole*

VOLUME 2

JOHN CORDY JEAFFRESON

CAMBRIDGE
UNIVERSITY PRESS

CAMBRIDGE
UNIVERSITY PRESS

University Printing House, Cambridge, CB2 8BS, United Kingdom

Published in the United States of America by Cambridge University Press, New York

Cambridge University Press is part of the University of Cambridge.

It furthers the University's mission by disseminating knowledge in the pursuit of
education, learning and research at the highest international levels of excellence.

www.cambridge.org
Information on this title: www.cambridge.org/9781108070751

© in this compilation Cambridge University Press 2014

This edition first published 1864
This digitally printed version 2014

ISBN 978-1-108-07075-1 Paperback

THE LIFE

OF

ROBERT STEPHENSON.

VOL. II.

LONDON
PRINTED BY SPOTTISWOODE AND CO.
NEW-STREET SQUARE

Engraved by Henry Adlard, from a.   Photograph by Mayer & Pierson, 1856

London Longman & Co

THE LIFE

OF

ROBERT STEPHENSON, F.R.S.

ETC. ETC.

LATE PRESIDENT OF THE INSTITUTION OF CIVIL ENGINEERS.

BY

J. C. JEAFFRESON

BARRISTER-AT-LAW

WITH DESCRIPTIVE CHAPTERS ON

SOME OF HIS MOST IMPORTANT PROFESSIONAL WORKS

BY

WILLIAM POLE, F.R.S.

MEMBER OF THE INSTITUTION OF CIVIL ENGINEERS.

IN TWO VOLUMES.

VOL. II.

LONDON:

LONGMAN, GREEN, LONGMAN, ROBERTS, & GREEN.

1864.

CONTENTS

OF

THE SECOND VOLUME.

———◆◇◆———

CHAPTER I.

THE BATTLE OF THE GAUGES.

(ÆTAT. 40–42.)

Great Western Railway in 1833—Brunel's Scheme for a Broad Gauge— History of the Narrow Gauge—Advantages anticipated by Brunel from a Broad Gauge—Brunel's Report of 1838—Theory of Railway Districts— Break of Gauge first takes place at Gloucester in 1844—Goods Traffic, not Passengers, the grand Cause of Difficulty at 'Breaks of Gauge'—Gauge Pamphleteers—The Oxford and Wolverhampton Contest in 1845—Lord Dalhousie's and Mr. Cobden's Motions—Royal Gauge Commission appointed—Brunel's Inconsistencies—The Railway Clearing House, instituted under the Auspices of Mr. Glyn and Mr. Hudson—Its leading Principles and its Returns for 1845—Witnesses examined by the Gauge Commissioners—Brunel left alone—Robert Stephenson's Character as a Parliamentary Witness—His Evidence before the Gauge Commission— Brunel's Expedients for obviating the Evils of 'Break of Gauge'—The Commissioners' Report—The last Argument in favour of Competition advanced by the Broad Gauge Party and answered by Mr. Thornton Hunt—Illustrated Evidence—The Gauge Act—Robert Stephenson's Report on Double Gauges   .   .   .   .   Page 1

CHAPTER II.

IRON BRIDGES.

Mr. Stephenson's large Practice in Iron Bridges—His Article on the Subject in the *Encyclopædia Britannica*—Modern Use of the Material— Early Bridges—First Iron Arch Bridges—Tom Paine's Bridge—Full

Development of the Iron Arch Bridge—First Use of Wrought Iron—Suspension Bridges—Captain Samuel Brown—Mr. Telford—The Menai Bridge—Introduction of Railways—Consequent large Demand for Iron Bridges—Return to the Form of the simple Beam—Comparison of the three different Systems of Iron Bridges—Advantages of the Girder System—Cast-iron Girders—Compound Girders—The Dee Bridge—Royal Commission on Iron Railway Structures—Introduction of Wrought-iron Girders—Different Varieties of Girders—Examples—The Aire Bridge—The Benha Bridge—Last Work of Mr. Stephenson's Life, Restoration of Tom Paine's Bridge . . . . . Page 30

CHAPTER III.

THE BRITANNIA BRIDGE.

The Port of Holyhead—The Holyhead Trunk Road—Interruption by the Menai Strait—Attempts to establish a Passage—Telford's Suspension Bridge—Introduction of Railways—Chester and Holyhead Railway—Proposal to use Telford's Bridge for Railway Purposes—Mr. Stephenson designs an independent Bridge—The Britannia Rock—Proposal for a Bridge of two Arches—Opposition in Parliament—First Idea of the Tubular Construction—Its Novelty—Preliminary Experiments : Mr. Fairbairn and Mr. Hodgkinson—Important Principles derived from the Experiments—Mr. Stephenson's Report—Commencement of the Masonry—Further experimental Inquiries—Means of placing the Tubes in their Positions—Contracts for the Tubes—Their Manufacture—Floating and Raising—Description of the Bridge—Principle of Continuity—Tubes—Mr. Stephenson's Explanations of Peculiarities in their Construction—Towers and Abutments—Architectural Design—Cost—The Conway Bridge 73

CHAPTER IV.

THE HIGH LEVEL BRIDGE AT NEWCASTLE-ON-TYNE.

Object of the Bridge—Ravine of the Tyne—Ancient Bridge at a Low Level—Inconveniences of the Passage—Early Proposals for a High Level Bridge—Mr. Green's Scheme—High Level Bridge Company—Mr. Stephenson appointed Engineer—Newcastle and Darlington Railway—Proposal for the Double Roadway—Parliamentary Proceedings—Description of the Bridge—The Piers—The Iron Superstructure—Mr. Stephenson's Motives for the Adoption of the Bowstring Girder—Letting of the Contract—Driving the Piles—Manufacture of the Ironwork—Erection—Completion 113

CHAPTER V.

AFFAIRS, PUBLIC AND PRIVATE, DURING THE CONSTRUCTION OF THE
CHESTER AND HOLYHEAD RAILWAY.

(ÆTAT. 42–47.)

Newcastle and Berwick Line—The High Level Bridge—Trent Valley
Line—Leeds and Bradford Line—Italian Trip in 1845—Norwegian Trip
in 1846—Norwegian Liberality—Irish Famine—Lord George Bentinck's
Proposal to. subsidise Irish Railway Companies—Robert Stephenson,
George Hudson, and Mr. Laing consulted—Lord George Bentinck's
Speech in the House of Commons—Election to the Council and Vice-
Presidency of the Institute of Civil Engineers—Narrow Escape on the
Chester and Holyhead Railway—Death of George Stephenson—Relations
between Father and Son—Elected Fellow of Royal Society—Grand
Banquet at Newcastle—Summary of his Railways—High Level Bridge
opened by the Queen—Robert Stephenson declines the Honour of Knight-
hood—'Nene Valley Drainage and Navigation Improvement Com-
missioners'—Appointed Engineer with Sir John Rennie to the 'Norfolk
Estuary Company'—Consulted by the Town Council of Liverpool as to
the best Means of supplying Liverpool with Water—Grand Central
Station at Newcastle opened by the Queen—Royal Border Bridge opened
by the Queen—Statistics relating to Royal Border Viaduct—Robert
Stephenson desirous of Rest . . . . Page 128

CHAPTER VI.

ROBERT STEPHENSON AS POLITICIAN AND MEMBER OF THE HOUSE OF
COMMONS.

(ÆTAT. 44–45.)

George Stephenson's Political Opinions and Sympathies—Robert Stephenson's Toryism—'*Little* Lord John!'—Opinions on Popular Education—
Robert Stephenson M.P. for Whitby in Yorkshire—'One of the Im-
penetrables'—Speech in the House of Commons on the proposed Site
for the Great Exhibition of 1851—Discussion on the Army Estimates,
June 19, 1856—First Speech against the Suez Canal—Second Speech
against the Suez Canal—Speech on 'The State of the Serpentine'—
Popularity in the House of Commons—Letter to Admiral Moorsom on
Crimean Mismanagement—Reason for declining the Invitation of the
Newcastle Conservatives—Dislike of Party Strife—Testimonial to Sir
William Hayter 142

CHAPTER VII.

ROBERT STEPHENSON IN LONDON SOCIETY.

(ÆTAT 47–55.)

The Year of the Great Exhibition—In the Park—An Impostor imposed
upon—No. 34 Gloucester Square—'The Sunday Lunches'—Works of

Art—Philosophical Apparatus—Demeanour in Society—'The Chief' in Great George Street—Robert Stephenson and 'the Profession'—Stories of Robert Stephenson's Generosity—'The Westminster Review' on Robert Stephenson—Cab-drivers and their Payment—Zenith of Robert Stephenson's Prosperity—His part in the Great Exhibition of 1851—Crack-brained Projectors—Aquatic Amusements—The 'House without a Knocker'—Alexandria and Cairo Railway—Victoria (St. Lawrence) Viaduct—Mr. Samuel Bidder's Reminiscences—Grand Banquet at Montreal to Robert Stephenson—His Speech on the Occasion—Connection with Mr. Alexander Ross—The St. Lawrence Viaduct completed, and inspected by Robert Stephenson's Deputies   .   .   Page 157

CHAPTER VIII.

THE GREAT VICTORIA BRIDGE OVER THE RIVER ST. LAWRENCE IN CANADA.

One of Mr. Stephenson's Last Works—Line of Lakes and River St. Lawrence—Difficulties of the Navigation—Introduction of Railways into Canada—The Grand Trunk Railway—Engineering Problem in the Design of the Bridge—Phenomena of the Ice—Early Proposals for a Bridge—Mr. A. W. Ross—Mr. Stephenson consulted—Joint Report—Mr. Stephenson visits Canada and reports again to the Directors—Surveys —Letting of the Contract—Iron-work—Mr. G. R. Stephenson—Controversy on the fitness of the Design—Mr. Stephenson's Views—Description of the Bridge—Site—Piers—Tubes—Erection of the Bridge—Foundations—Caissons—Shortness of the Working Season—Contrivances to save Time—Inspection of the Bridge—Opening by the Prince of Wales—Difficulties overcome   190

CHAPTER IX.

CONCLUDING YEARS AT HOME AND ABROAD.

(ÆTAT. 47-55.)

Athenæum Club—Geographical Society—Royal Society Club—The 'Philosophical Club' of the Royal Society—Robert Stephenson, President of the Institution of Civil Engineers—Receives the honorary D.C.L. of Oxford—The Dark Side of his Prosperity—Failing Health—Admiration of Mechanical Skill—Speech at Sunderland—The Wear Bridge—Strong Affection for Newcastle—Periodic Visits to the Factory—Judicious and Considerate Conduct to humble Relations—Visits to Long Benton—Contribution to Painted Window in Long Benton Church—Visit to Wylam—Isaac Jackson, the Clockmaker—On Board the 'Titania'—Letter to Admiral Moorsom—Hampstead Churchyard—Social Engagements—Trip to Egypt—An unfulfilled Presentiment—Letters from Alexandria and Algiers—Last Christmas Dinner of Stephenson and Brunel—Last London Season—Last Visit to Royal Society Club—Last Will and Testament—Last Voyage to Norway—Opening of the Norwegian Railway —Banquet to Robert Stephenson at Christiania—Last Public Speech. 230

CHAPTER X.

LAST SCENES.

(ÆTAT. 55.)

Homeward Course of the 'Titania' and the 'Mayfly'—Robert Stephenson lands at Lowestoft—Arrives at Gloucester Square again—Temporary Rally—Death—Public Agitation—The Queen's Expression of Sympathy —Funeral Procession—Interment in Westminster Abbey—Attendance at the Ceremony—Sacred Service at Newcastle—Public Mourning at various important Towns—Plate on Coffin-lid—Inscription on Monumental Brass —The Article in the 'Times' on the Morning after the Funeral— Generous Tone of the Press—Last Honours . . Page 258

INDEX 306

ILLUSTRATIONS IN VOL. II.

—◆—

PORTRAIT OF ROBERT STEPHENSON, FROM A PHOTOGRAPH   *To face Title*

DIAGRAMS OF SEVERAL LARGE IRON BRIDGES, ALL DRAWN TO

THE SAME SCALE . . . . *Page* 71

BRITANNIA BRIDGE, NORTH WALES . . . . . „ 73

CONWAY BRIDGE, NORTH WALES . . . . . . „ 109

HIGH LEVEL BRIDGE, NEWCASTLE-UPON-TYNE . . . „ 127

VICTORIA BRIDGE, OVER THE ST. LAWRENCE, MONTREAL . „ 227

THE LIFE

OF

ROBERT STEPHENSON.

————∘∘⚬✿⚬∘∘————

CHAPTER I.

THE BATTLE OF THE GAUGES.

(ÆTAT. 40–42.)

Great Western Railway in 1833 — Brunel's Scheme for a Broad
Gauge — History of the Narrow Gauge — Advantages anticipated
by Brunel from a Broad Gauge — Brunel's Report of 1838 — Theory
of Railway Districts—Break of Gauge first takes place at Gloucester
in 1844—Goods Traffic, not Passengers, the grand Cause of Difficulty
at 'Breaks of Gauge'—Gauge Pamphleteers—The Oxford and
Wolverhampton Contest in 1845—Lord Dalhousie's and Mr. Cobden's
Motions—Royal Gauge Commission appointed—Brunel's Inconsistencies—The Railway Clearing House, instituted under the
Auspices of Mr. Glyn and Mr. Hudson—Its leading Principles and
its Returns for 1845—Witnesses examined by the Gauge Commissioners—Brunel left alone—Robert Stephenson's Character as a
Parliamentary Witness—His Evidence before the Gauge Commission—Brunel's Expedients for obviating the Evils of 'Break of
Gauge'—The Commissioners' Report—The last Argument in favour
of Competition advanced by the Broad Gauge Party and answered
by Mr. Thornton Hunt—Illustrated Evidence—The Gauge Act —
Robert Stephenson's Report on Double Gauges.

WHILST Robert Stephenson was proving that the
locomotive was superior to atmospheric propulsion in economy and adaptability of power, he was

involved in another controversy, of not less importance, which brought him again in collision with the brilliant engineer, who was throughout life his constant professional opponent, and warm private friend. The relations that subsisted between him and Brunel could not have endured between rivals endowed with merely ordinary generosity. Continually as they were pitted against each other, much as the reputation of the one was exalted by the failures of the other, they not only preserved strong mutual affection, but in their gravest periods of public trial were always ready to assist each other with counsel and support. When Robert Stephenson with fearful anxiety was watching the floating of his first enormous tubular bridge to the piers, Brunel stood by his side ; and when Brunel was heroically contending with the gigantic difficulties of launching the Great Eastern, Robert Stephenson disregarded the claims of failing health, in order that he might be on the spot to encourage and advise his brother engineer. Two nobler adversaries the world never witnessed.

Whilst ordinary men were admiring the phenomena of railway developement, Brunel was criticising George Stephenson's system and planning improvements. It struck him that iron roads were not all they might be, or ought to be ; and it was not long before he struck out a novel method for their construction. At the first projection of the Great Western Railway in 1833, it was contemplated that that line and the London and Birmingham Railway should have a common terminus in the metropolis. The combined opposition of the Eton and Oxford authorities threw out the Great Western Bill in its first parliamentary campaign, and before the renewal of the contest, Brunel,

as engineer of the line, proposed to some of the directors
that their gauge, or distance between the rails, should be
7 feet instead of 4 feet 8½ inches. This suggestion was
submitted to Robert Stephenson, and was by him promptly
rejected. Under ordinary circumstances there would
have been an end of the novel scheme ; but Brunel was
gifted with no ordinary powers of persuasion, and the
directors of the Great Western were induced by him
to separate themselves from the London and Birmingham Company, and make their line according to his
wishes.

As the reader is well aware, the gauge of George
Stephenson's first public railway was 4 feet 8½ inches,
which had been the gauge of the colliery tramways of
Northumbria from the time of their first construction.
In the Life of Lord Keeper North, A.D. 1676, it is
recorded —

The manner of the carriage is by laying rails of timber
from the colliery to the river, exactly straight and parallel ; and
bulky carts are made with four rollers fitting those rails, where-
by the carriage is so easy that one horse will draw down four or
five chaldrons of coals, and is an immense benefit to the coal
merchant.

Made to be drawn by horses, these wagons differed
little from the carts previously used, the innovation
consisting only in finding for them smooth wooden ways,
and wheels adapted to those ways. When the wooden
trams were first cased with metal, and later on the introduction of iron rails, the same width was continued. The
introduction of the locomotive brought with it no new
conditions inviting men to change the usage of the
country; and George Stephenson therefore made his

lines in accordance with the ancient custom. This 4 foot 8½ inches was the original tramroad gauge.

Other gauges were in existence. In some of the mineral districts of England, where the tramways have to meander down hills and into positive gullies, a gauge of two feet had been adopted. In such a country, and for the carriage of minerals, a very broad gauge was clearly not to be thought of. But for comparatively open and level regions no objection to the introduction of greater width between the rails presented itself, to counterbalance the advantages hoped for from the change. Those advantages Brunel expected to find in greater speed, ease of motion, and economy of working. With the wider way, the engineer contemplated the use of larger carriages and more powerful engines. From his engines fitted with wheels ten feet high he looked for a vast increase of speed; and he hoped to effect greater safety by placing his passenger-carriages between instead of over the wheels. According to his calculation one grand advantage of the wide gauge would be diminution of oscillation at high speeds.

The most obvious objection to a wider gauge, at that period of railway history, was the increase it would necessarily effect in the expenses of constructing a line—especially where tunnels, earth-works, and viaducts were frequently needed. The next point for criticism to fix upon was the inconvenience that would ensue to the public wherever lines with different gauges ran into each other. These two difficulties Brunel handled with characteristic adroitness, treating the former as of little weight with regard to the works he contemplated, and finding in the latter an argument actually in favour of his scheme.

Making the most of his theory that each district of the
country should have the gauge most adapted to its geographical features, he reminded his opponents that it was
no part of his plan to do away with the two, three, and
four feet gauges of mineral districts, or to oppose the
4 feet 8½ gauge in countries where that width had already
been used or was likely to be most serviceable, but only to
introduce his wide gauge in regions, comparatively open,
sparsely populated, and untried by railway engineers.
London and Bristol, he argued, were separated by a sweep
of country offering (except at two or three points) comparatively few obstacles to the maker of iron roads.   The
difference of cost, therefore, between a wide road and a
narrow road would be slight—at least slight compared
with the advantages of a system which would convey with
unexampled rapidity an entire army of passengers from
the metropolis to the capital of the West, in a single
train.   So cleverly was the objection of expense thus put
aside, that shareholders were almost ashamed of their
folly in raising the question.   The next point—the inconvenience, namely, of 'break of gauge,' as it was soon
called—Brunel treated in a very different way.   It was
true the inconvenience of a break of gauge would be
grave, if it occurred; but then he maintained it never
would occur.

In his report of 1838 to the directors of the Great
Western, he said:—

I shall now consider the subject of the width of gauge.   The
question of the disadvantage of differing in point of gauge from
other railways, and the consequent exclusion from communication with them, is the first.   This is undoubtedly an inconvenience; it amounts to a prohibition to almost any railway

running northwards from London, as they must all more or less depend for their supply upon other lines or districts where railways already exist, and with which they most hope to be connected. In such cases there is no alternative.

The Great Western Railway, however, broke ground in an entirely new district, in which railways were unknown. At present it commands this district, and has already sent forth branches which embrace nearly all that can belong to it.

Such is the position of the Great Western Railway. It could have no connection with any other of the main lines, and the principal branches likely to be made were well considered, and almost formed part of the original plan, nor can these be dependent upon any other existing lines for the traffic which they will bring to the main trunk.

Such was Brunel's language in the early stages of the gauge controversy, and such it had been when he prevailed on the directors of the Great Western to adopt his innovation.* Briefly stated, his argument was this :—

---

* Mr. Brunel's evidence before Gauge Commissioners, Oct. 25, 1845, gives the particulars of the origin and growth of his preference for the Broad Gauge.

'You are the engineer of the Great Western Railway ?—I am.

' Was the line surveyed under your direction ?—Yes.

' And you decided on its course ? —Yes.

' In what year was that ?—In 1833.

' That was three or four years subsequently to the formation of the Manchester and Liverpool Railway ?—Yes.

' Had you, before you took the direction of the Great Western Railway, any employment in railway matters ?—No.

' That was the first line upon which you were engaged as an engineer ?—Yes, the first line upon which I was engaged which was constructed ; I had looked over other lines of country.

' With a view to railways ? — Yes.

' At what period did it occur to you to change the gauge from 4 feet 8½ inches to 7 feet ? — I think, in my own mind, it occurred to me in the course of my surveys in 1833 and 1834.

' That a change of gauge would be desirable ? — Yes.

' But the exact amount of the change you had not then decided upon ? — I think not, and I think I never mentioned it to anyone.

' Will you favour the Commissioners with the reasons which induced you to think that 4 feet 8½ inches was insufficient at that early period ? —*Looking to the speeds which*

The west country at present has no railways, it lies open to our enterprise. Let us seize the opportunity, and drive a grand trunk line with a few important branches through it, making our gauge such that no line of the old gauge can run into our roads and suck our traffic. By adopting this course we shall have a monopoly of the west country.

At first Brunel met with little encouragement from the directors. They were not alarmed at the novel proposals, nor did they condemn them as chimerical, but commercial caution made them apprehensive that they might sink in public estimation if they declared themselves the leaders of a revolutionary movement. Brunel's suggestion, however, of a monopoly of the west country, from the impossibility of narrow gauge lines acting harmoniously

---

I contemplated would be adopted on railways, and the masses to be moved, it seemed to me that the whole machine was too small for the work to be done, and that it required that the .parts should be on a scale more commensurate with the mass and the velocity to be attained.

'The trains at that period were comparatively light to what they are now, both in goods and passengers? —Yes.

'You had probably travelled a good deal upon other railways, and had seen much of other railways that then existed? — Yes, as much as I possibly could. I think the impression grew upon me gradually, so that it is difficult to fix the time when I first thought a wide gauge desirable; but I dare say there were stages between wishing that it could be so, and determining to try and do it, and I cannot at this moment distinctly remember the time.

'Do you recollect at what period you determined upon submitting the 7 feet gauge to the directors of your company? — It must have been almost immediately after the passing of the Act, which was in 1835, and I think I must have mentioned it to the directors long prior to that, because I made great efforts to get the clause omitted which fixed the gauge, and I communicated certainly with Lord Shaftesbury early in 1835.

'Therefore the omission of that clause, which was a very proper omission perhaps, was the result of your communication? — It had been omitted, fortunately perhaps for me, in one Bill previously. I think that the Commissioners will find that in the first Southampton Railway Act it was omitted. It was omitted in the first Great Western Bill, and there I must have taken steps with reference to the gauge early in 1835.'

with broad gauge lines, sunk deep in the minds of the projectors, and bore fruit.

There is no ground for thinking that Brunel acted disingenuously towards his directors. He saw in railways only the future channels of communication between important centres of manufacture and commerce—not the means of passage between petty market towns and secluded hamlets. Each range of country would have its grand trunk, with its limited number of branches to cathedral towns and harbours; but it was not on the list of chances that the branches of these gigantic arteries would multiply, extend, and cross each other—that the surface of the island would be one patch of network. Holding this view (which was the view almost universal in 1833), Brunel gave his directors honest counsel.

He gained his object. The bill was obtained, and the line was made in accordance with his wishes. It was true that its construction was attended with costly accidents and vain experiments. The engines with the huge wheels turned out failures, in consequence of their being deficient in boiler power; but at length the railroad began its career with dazzling *éclat.* The Great Western was the topic of ' the season.' Everyone was in raptures with the smoothness of its way, the height of its speeds, and the luxury of its first-class carriages. As far as the drawing-rooms of May Fair were concerned, the success of the broad gauge was established. Many a humble family has cause to lament that experience, and vulgar calculations of pounds, shillings, and pence have signally falsified this flattering verdict.

A few years gave the public an opportunity of judging how far the theory of distinct fields of railways, not running into each other, was likely to be realised in practice. The plans of projectors soon indicated that iron roads would refuse to run to the capital without inter-communication, and the year 1844 saw the Western and Midland counties in actual collision. The extension of the line between Birmingham and Gloucester, uniting the latter town with Bristol, had, in order that it might accord with the line of which it was a continuation, been planned on the narrow gauge. The directors of the Great Western, seeing in this narrow gauge extension, known as the Bristol and Gloucester, an alarming irruption into their broad gauge field, contrived by a stroke of finance to gain control over its company. Their control was of course exercised to convert the proposed narrow gauge into an actual broad gauge. The result was that on the opening of the extension in 1844, the two scales of road-way met, and Gloucester had the honour of being the scene of the first 'break of gauge.' At first ' the break ' attracted but little attention beyond engineering circles. The public were not sufficiently familiar with railways to be highly critical. If passengers from Birmingham to Bristol had to get out of narrow gauge carriages at Gloucester, and crossing over a platform with their baggage, had to seek fresh places in the broad gauge extension, the trouble was trifling compared with that of the shiftings from stage-coach to stage-coach to which travellers had been accustomed. When ' the battle of the gauges was at its height,' pamphleteers were pathetic on the sufferings of delicate ladies and young children, compelled to ' change places,' and pass through the raw night air on their way from one gauge to the other.

Had passengers only been affected by 'break of gauge,' little attention would have been paid to their discomfort and complaints; for the hardship is slight which an ordinary traveller sustains in changing his carriage once in half a hundred miles. The real inconvenience of 'a break of gauge' was found in the conveyance of goods.

Railway communication had not existed many weeks between Birmingham and Bristol, before the manufacturers of Birmingham and the railway officials at Gloucester knew what was the real difficulty. The heavy goods, sent from Birmingham for shipment at Bristol, had to be shifted from gauge to gauge by the Gloucester porters. Packages were misplaced, delayed, or missent. Complaints daily increased; and 'Birmingham men' learnt the discomfort of having a break of gauge between themselves and the Bristol Channel. In due course a comparison of the goods traffic on the Grand Junction, the London and Birmingham, and the Midland lines, with that on the route between Birmingham and Bristol, gave a triumph to the opponents of the broad gauge.

'Break of gauge' was no longer a matter of speculation, but an evil in actual existence. The agitation it aroused soon attracted the attention of the legislature. In the session of 1845, the London and Birmingham and Great Western Companies were in the field with rival bills for a line of railway between Oxford and Wolverhampton. The manifest evils of 'break of gauge' induced the railway department of the Board of Trade to decide against the pretensions of the Great Western. The House of Commons, however, set aside the decision of the Board of

Trade, and without offering any opinion on the advantages of uniformity or variety of gauges, gave the preference to the Great Western on the ground that it was the better line, their choice being endorsed by the House of Lords. Thus for the moment victory was with the broad gauge party, but the facts brought to light by the contest between Robert Stephenson's company and Brunel's company induced both Houses of Parliament to ask for further investigations.

The battle now began in earnest. All the preceding encounters were mere skirmishing, compared with the tug of war which now set in. On the one side were drawn up the forces of narrow gauge, on the other appeared those of broad gauge, double gauge, and mixed gauge; whilst hovering on the flanks of the two armies were the scattered companies of the medium gauges.

The revelations of the Oxford and Wolverhampton Committee were followed by the motions of Lord Dalhousie in the Lords and Mr. Cobden in the Commons, which resulted in an address, unanimously voted, for a Royal Commission to ascertain ' whether in future private acts for the construction of railways, provision ought to be made for securing an uniform gauge; and whether it would be expedient and practicable to take measures to bring railways already constructed, or in progress of construction, into uniformity of gauge.' Without delay the commission was appointed. It was composed of Colonel Sir Frederick Smith, of the Royal Engineers, who had previously acted as Inspector-General of Railways; Professor Barlow, of the Woolwich Military Academy (ex-commissioner of Irish Railways), and Professor Airy, the Astronomer Royal.

The men whose memories survey seven years with accuracy are few. But the few who bore in mind Brunel's position in 1836, '37, and '38 were not a little amused with the line he adopted before the Oxford and Wolverhampton, and the Gauge Commissions. Of course, inconsistencies and plausible arguments in their support were looked for from the man who, in the full observation of men of science and the general public, appeared as the champion of two novelties in railway locomotion, involving diametrically opposite principles. While the broad gauge demanded larger and heavier locomotives than the narrow gauge, the atmospheric system was represented as immeasurably superior to the locomotive system, because the grinding weight of the travelling engines (such as were used on narrow gauge lines) caused ruinous damage to the rails.. That is to say, Brunel, at one and the same time, was exclaiming against the destruction of rails by the use of heavy locomotives, and urging the employment of locomotives of an unprecedented weight. On the broad gauge Brunel asked for easy curves; while he represented that the peculiar merit of the atmospheric system was the capability which it afforded of constructing lines with very sharp curves. In railway administration, argued the engineer of the broad gauge lines, the first object was to limit the traffic to a few heavy trains; but changing his tone, the versatile engineer, in pleading for the atmospheric system, insisted that the exigencies of the public required trains to be many and light.

Nothing daunted Brunel. His theory of railway districts had signally broken down. At the outset of his crusade against the narrow gauge, he had argued that

break of gauge could never happen — partly because
railway lines would have a natural tendency towards
London, and partly because the enormous and manifest
inconveniences of 'a break of gauge' would deter any
line of one width from running into another. But the
break, which he maintained sheer terrorism would
render an impossibility, had through his instrumentality
occurred. The case was unquestionably an awkward one,
but he could meet it. He stood up, and smiled at the
fears of his opponents. It was true that breaks of
gauge, if the broad gauge system were extended, would be
frequent, * but they could be easily dealt with. With

---

* 'The completed or projected
branches of the Great Western
Railway itself,' says Mr. Wyndham
Harding in his pamphlet — 'The
Gauge Question; Evilsof Diversity of
Gauge, and a Remedy '—' which was
expected, as we have seen, to have no
connection with any other existing
line—now join it to most of the other
main lines in the country. For
instance:

' To the Grand Junction, and to the
projected Shrewsbury and Birming
ham Railways, at Wolverhampton.

' To the Grand Junction, London
and Birmingham, and Midland Railways, at Birmingham.

' To the London and Birmingham,
the Midland, and the proposed Trent
Valley and Churnet Valley lines, at
Rugby.

'To the London and Birmingham
Railway again, at Warwick.

' To the Birmingham and Gloucester Railway, at Cheltenham and
Worcester.

' To the South Western Railway,
at Basingstoke and Salisbury.

' To the projected Dorchester and
Southampton Railway, at Dorchester.

' To the proposed Welsh Midland
line, at Hereford and Swansea.

' To the Bristol and Gloucester
line, with which it is already
connected, at Bristol and Stonehouse.

' [All these are narrow gauge
lines, with the exception of the last,
which is a broad gauge line at
present, but its proprietors have announced their desire and intention of
obtaining powers to convert it into a
narrow gauge line.]

' And if the Great Western Railway, with its broad gauge branches,
do not go to these lines, they with
their narrow gauge branches will come
to the Great Western, thus connecting by railway almost every county
and town in the kingdom with every
other.

' What are all these branches projected for, except to bring traffic from

inexhaustible fertility of resource, he enumerated various expedients by which the gigantic evil of 1838 could in 1845 be reduced to a merely nominal inconvenience. The passengers could be left to take care of themselves. As for goods, porters could shift small packages by hand. Heavier goods might be packed on carriages, so constructed that their bodies by the aid of a mechanical apparatus might be shifted, without unpacking their contents, from frames with narrow wheels and axles to broader frames with wheels and axles suited to the wide gauge. He was even prepared to shift the narrow gauge carriages, wheels and all, and place them on broad gauge frames. Coals and other minerals might be packed in loose boxes made of iron, two of which when shifted from the narrow roads would fill one broad gauge truck. He invented telescopic axles which enabled carriages to travel on either gauge. Or he would lay down narrow gauge lines within his broad gauge rails. Of course these expedients involved great additional expense, in porters and machinery and time. But pecuniary expense was a consideration to which Brunel was indifferent.

A few extracts from his evidence before the Gauge Commissioners will show how little care he now had for expenditure in an arrangement which was in the first instance recommended to the public on the score of economy.

---

the lines and districts with which they communicate, or to take traffic to them from one extremity of the country to another, and therefore over the narrow gauge on to the broad gauge, or over the broad gauge to a narrow gauge? The diffi-culties attending a change of gauge then, which, as was admitted, would in 1838 " have entirely prevented in the north such a course " as one railway adopting different dimensions from the rest, now have " existence in the west." '

As regards goods (he said in the conclusion of his long answer to interrogatory 4029), *it is of course a mere question of money*; and if there is a considerable stream of goods in one line, and it is the interest of two parties meeting at a certain point to interchange those goods, I believe the inconvenience *and expense will be so trifling that it is hardly worth consideration*, if there are other important considerations in the question of the change of gauge.

When, however, he was pressed as to the details of his plans for interchange, he became even more vague :—

4048. Having dealt with the passengers, and having had now some considerable time to think of the question of goods, since it was brought forward in the last session of parliament, have *you made up your mind at all as to the mode in which you would arrange respecting them?*—No; because it must depend upon what other companies choose to do on the other side; if they do not afford assistance, I will not say if they throw impediments in the way, but if they do not afford assistance to exchange, the mode must be different from that which it would be if they did. As regards coal, there is no doubt that there would be every facility, because the mode of carrying an article in large quantities like coal, will no doubt be influenced by the wishes and desires of the coal-owners, and the coal-owners will, of course, be desirous of doing whatever will encourage their trade with Oxford.

4049. You would have no difficulty with them ?—I think we shall have no difficulty whatever with them. As regards general goods, it must depend upon what the other companies may choose to do; the worst that could happen, of course, would be the entire unloading and reloading of goods ; *even that does not amount to anything in time or money that would be much felt by the public.*

A reference to the returns for the year 1845 of the Railway Clearing House will show how far interchange between railways could be hindered or disturbed, without causing the public serious inconvenience. Under the auspices of Mr. Hudson and Mr. Glyn, the Railway Clearing House was established on January 2, 1842, to relieve railway companies of the burdensome calculations consequent on a system of correspondence which had grown up since the opening of the public lines of railway between London and Liverpool. The leading principles of the Clearing House system are three.* *Firstly*, passengers are booked through at all principal stations, and conveyed to their destinations without change of carriage—horses, cattle, goods, being in like manner sent through without change of conveyance. *Secondly*, companies respectively pay a fixed rate per mile, for such carriages and wagons, not their own property, as they may use ; and a further sum per day by way of fine or demurrage for detention, if kept beyond a prescribed length of time. *Thirdly*, no direct settlement may take place between the companies in respect of any traffic, the accounts of which have passed through the Railway Clearing House. This is no place for a minute description of the Clearing House operations. It is enough to say, that through them ' the transactions of one company with all other companies, amounting frequently to many thousand pounds a week, are cleared weekly by a sum seldom exceeding a few hundred pounds.'

An institution so manifestly adapted to commercial exigencies met with immediate success. With a few very unimportant exceptions, all the narrow gauge companies joined the association as soon as they came into existence, and had need to correspond with other companies. The

---

* 'The Origin and Results of the Clearing System which is in operation on the Narrow Gauge Railways.'   A pamphlet printed for private circulation.

transactions of the House soon became very heavy; their returns for 1845 showing,

that 517,888 persons were in that year each conveyed through an average distance of 146 miles, the average length of the lines travelled over being forty-one miles, so that each passenger travelled over four railways on the average, and must have passed three junctions or points of convergence. To accommodate these passengers, 59,765 railway carriages, and 5,813 carriages, were sent through. There were also sent through in the same year, 7,573 horse-boxes, 2,607 post offices, and 180,606 goods wagons, besides wagons conveying minerals, of which no record is kept in the Clearing House.*

Of course the extension of the broad gauge would have crippled, if not altogether put an end to this admirable system of correspondence. The Clearing House, therefore, was another powerful antagonist to the broad gauge, and its returns, giving the aggregate of railway correspondence throughout the entire country, furnished the advocates of uniformity of gauge with valuable facts and illustrations for their arguments.

The witnesses examined by the Gauge Commissioners in 1845 were forty-six † in number, and they included every

---

* 'A Brief History of the Gauge Question.'

† The following classified list of the forty-six witnesses will be interesting both to the public and to professional readers:—

In favour of uniformity and a narrow gauge:

1. Bass, William, agent to Messrs. Pickford.
2. Bidder, George Parker, C.E.
3. Bodmer, George, locomotive manufacturer.

4. Braithwaite, John, C.E.
5. Brown, James, manager of Sir John Price's iron and coal-works.
6. Buckton, Thomas, secretary to the Brighton Railway.
7. Budd, James P., manager of copper-works and coal-mines, and deputy-chairman of the Welsh Midland Railway.
8. Chaplin, W. James, chairman of the South Western, head of the carrying firm.
9. Clarke, Peter, manager of the Brighton Railway.

person eminent in the railway world as an engineer, a manufacturer of locomotives, a manager, a secretary, a

10. Creed, Richard, secretary to the London and Birmingham Railway.

11. Ellis, John, deputy-chairman of the Midland Railway.

12. Fernihough, William, locomotive superintendent of Eastern Counties Railway.

13. Harding, Wyndham, late manager of the Bristol and Gloucester Railway.

14. Hawkshaw, John, C.E.

15. Hayward, Joseph, of the firm of Pickfords', carriers.

16. Horne, Benjamin W., carrier.

17. Hudson, George, M.P.

18. Huish, Mark, Capt., general manager of the Grand Junction, and Liverpool and Manchester railways.

19. Jones, Evan, agent for Chaplin and Horne, carriers, at Camden Station.

20. Laws, R.N., Capt. J. M., general manager of the Leeds and Manchester Railway.

21. Locke, Joseph, C.E.

22. M'Connell, James Edward, superintendent of the locomotive department on the Birmingham and Bristol Railway.

23. Martin, Albinus, C.E., resident engineer and superintendent of the South Western Railway.

24. Mills, T. C., manager of the goods department of the London and Birmingham Railway.

25. O'Brien, Capt. William, late secretary to the South Eastern Railway.

26. Rastrick, J. U., C.E.

27. Stephenson, Robert, C.E.

28. Whitaker, Thomas, C.E.

29. Woods, Edward, C.E.

30. Wood, Nicholas, C.E.

In favour of an intermediate gauge, theoretically, and against the broad gauge, but favourable to uniformity of gauge:

1. Bury, Edward, locomotive manufacturer.

2. Gray, John, locomotive superintendent, Brighton Railway.

3. Pasley, Major-General, R.E., inspector of railways.

4. Roberts, Richard, locomotive manufacturer.

5. Vignoles, C., C.E.

In favour of an intermediate gauge, expressing no decided opinion as to uniformity of gauge:

1. Cubitt, Benj., late locomotive superintendent of the Croydon and South Eastern Railways.

2. Cubitt, William, C.E.

3. Landmann, Col. R.E.

Opposed to break of gauge, but expressing no opinion about gauge:

1. Burgoyne, Major-General Sir John, Quarter-Master General.

2. Gordon, General Sir Willoughby, Quarter-Master General.

3. Downs, Richard, contractor.

4. Jackson, Thomas, ditto.

In favour of broad gauge, and against uniformity:

1. Brunel, Isambard Kingdom, C.E.

2. Clark, Seymour, superintendent of traffic on Great Western.

3. Gooch, Daniel, superintendent of locomotives on Great Western.

4. Saunders, Charles Alexander, secretary of the Great Western.

carrier, or an amalgamator. Of them, only four were in
favour of a seven-foot gauge and against uniformity. Three
offered no opinion as to the desirability of a uniform
gauge. But all the others—i. e. 39 out of 46—were so
impressed with the inevitable evil consequences of break of
gauge, that they concurred in desiring uniformity of road
width, though five of that number had a theoretical
preference for an intermediate gauge, and four others
refrained from offering an opinion as to which gauge
was best.

Brunel found himself alone. Not one member of his
profession sided with him. Indeed his only companions
in 'the forlorn hope,' of which he was the leader, were
three gentlemen holding office under the Great Western
Company, and pledged in honour to fight to the death for
the broad gauge. At this date it is easy to see how
Brunel fell into his error, but it is difficult to judge him
with the generosity he merits. He was betrayed into an
embarrassing position not so much by seeing less, as by
seeing farther, than ordinary men. Of the crowd of
witnesses who came against him in 1845, there were few
who in 1834 and 1835 thought of the consequences of
'a break of gauge.' He, however, foresaw them, and
fancied that by seizing a wide tract of country he could by
the fear of those very consequences drive off competition.
Had he been a few years sooner in the field, he might
possibly have succeeded so far as to make his gauge the
gauge of the southern districts of the country, and in some
counties not only to have checked, but even to have sup-
planted the narrow gauge. When witness after witness
came up to beat down his fallacies before the Gauge Commissioners, they were only proving to him what he had

seen ten years before, and they had only learnt by recent experience. It is true that Robert Stephenson had seen farther than Brunel. He had not only foreseen the evils that would arise from 'a break of gauge,' but with his clear vision, and thorough familiarity with the powers which he and his rival were contemplating, discerned that in spite of those obstacles, different fields of railway would run into each other.

As Robert Stephenson had been the first to foresee the evil consequences of diversity of gauge, he was, apart from being the recognised chief of his profession, selected as leader of the narrow gauge party before the Commissioners. In this position, therefore, he was the first to give his evidence before the Commission on August 6, 1845. On all occasions Robert Stephenson's evidence was peculiarly impressive. If he saw the truth, he stated it, although it was against his interests. With equal honesty, he declined answering a question of opinion, if he had not sound and valid reasons wherewith to support his reply, even when he might feel confident that a quick off-hand statement, agreeable to his interests, would gain a point. 'I cannot answer that question,' was often heard from his lips. This straightforward candour at first told against him, but the influence of his testimony was in the long run enormously enhanced by it. When it was stated in railway discussions, that ' Robert Stephenson *said* such or such a thing,' it was understood that the statement, be it right or wrong, was the conscientious and deliberate opinion of the first practical engineer of his day.

Before the Gauge Commission, Robert Stephenson's evidence was temperate and convincing, as it was whenever he spoke on a subject connected with his profession. At the outset he admitted that at one time the narrow gauge had appeared to him too confined; and then he succinctly stated the changes which had removed the considerations on which that opinion was based.

As an engine builder (he said), at one time when I was called upon to construct engines of greater power than we commenced the line with, I felt some inconvenience in arranging the machinery properly; we were a little confined in space, and at that time an increase of three or four inches would have assisted us materially, and to that extent I thought at one time that an addition to five feet would have been desirable, but on no other account, looking at it as a mere engine-builder. Since that time the improved arrangements in the mechanism of the locomotive engines have rendered even that increase altogether unnecessary; at present, with the inside cylinders, which is the class of engine requiring the most room between the rails, and the cranked axle with the four eccentrics, we have ample space, and even space to spare.

With reference to space, in the arrangement of the machinery, which is the main question having reference to the width, the working gear has been much simplified, and the communications in the most recent engines, between the eccentric and the slide valve, have been made direct communications; whereas formerly it was made through the intervention of a series of levers, which occupied the width. But even without that which I have just now alluded to, which gives us an extra space with the engines on the South Western and on various lines in this country by the improvements which have been made, there is quite space enough for the whole of the working gear.

Then with reference to the increase of power, the size of boiler is, in point of fact, the only limit to the power, and we have increased them in length on the narrow gauge, because we have always made the boiler as wide as the narrow gauge would admit of, but we have increased their length both in the firebox and in the tubes; we have obtained economy, I conceive, by lengthening the tubes, and we have obtained an increased power by increasing the size of the fire-box; in fact, the power of the engine, supposing the power to be absorbed, may be taken to be directly as the area of the fire-grate, or the quantity of fuel contained in the fire-box.

After enumerating the different items—roadway, tunnels, embankments, viaducts, bridges—which would absorb the funds of a broad gauge company in the process of construction, he took into consideration the various expedients suggested by Brunel for effecting transfer of goods. The loose-box system, experience, he said, had proved to be a failure.

Whilst I think the Great Western has obtained no advantages by the wide gauge, I think its introduction has involved the country in very great inconvenience, because wherever a meeting of the gauges takes place, it must create an inconvenience, and a very serious one; in fact, it is nothing more or less than tantamount to asking the Great Western or the London and Birmingham Company to move their passengers at Wolverton; that is an exaggerated case perhaps, but still it is one which, if it takes place in the midst of a large traffic, would, I believe, give canals or another existing mode of communication a decided advantage over a railway. I stated in my evidence before the Wolverhampton Committee, that from Rugby, to which point it is proposed that the wide gauge should come, the Derbyshire coal-owner, or the Leicestershire, would inevitably send their coal by canal, in preference to changing the gauge, because they would have to transfer their coals there; it is proposed in order to avoid the actual removal of the coals, to move them in boxes, and to have loose bodies to the wagons. Now, that is a system which has been tried over and over again, and which has failed. It was tried on the Liverpool and Manchester line originally. There was a great coal-pit about 200 or 300 yards from the line of railway; they wanted to send coals to Liverpool, and small wagons were placed on the backs of large wagons, and carried to Liverpool; that was soon abandoned. Loose boxes were tried at Bolton for the purpose of leading the coal into the town by horses, without changing at the station; they were eventually abandoned. I tried the same thing at Canterbury, and we were obliged to abandon it, because sometimes we had loose boxes and we had no frames, and sometimes we had underframes when we had no boxes, and we could not fit them in. It is almost impossible to make this intelligible to any one who has not come directly in contact with the inconvenience of the system. Rather than introduce the loose-box system, it would be far better to move the coals by hand from wagon to wagon, because there would be an end of it. It also involves this, which I felt particularly at Canterbury: when the body of the wagon is attached, and made part of, and formed at the same time with, the frame, it strengthens that frame, and it strengthens also the body itself; but when they are made to separate, they are both of them weak, and they both get rickety, and they are exceedingly costly to maintain in repair.

The continuation of this portion of his evidence went to show that the evils enlarged upon were not so much defects from want of good management, but defects from which the system could not be freed.

As to the expedients proposed for shifting bodies of wagons by machinery, he showed that the time required for their application would render them commercially impracticable.

With similar force the witness unfolded the objections to the double gauge, i. e. the system which employs both gauges, by putting down narrow gauge rails within the broad gauge; and to the mixed gauge, or system which accommodates both broad and narrow carriages, by putting within the broad gauge a single rail at such a distance from the outer rail of the broad gauge that a narrow gauge line is thereby formed.

His evidence made it clear that uniformity of gauge was imperatively demanded for the transaction of business; that the expedients proposed for overcoming the discomforts of ' break of gauge ' would scarcely mitigate them ;

and that since an uniform gauge throughout the country was required, no gauge was so well adapted as the 4 feet $8\frac{1}{2}$ inches for all varieties of country.

Uniformity of gauge being the grand object, the following table* of the lines completed, in progress, and projected, in 1845, will show how strong a case the advocates of the narrow gauge had for maintaining that in regard to comparative interests involved, apart from all other considerations, the preference ought to be given to their system.

NARROW GAUGE RAILWAYS.	Miles	BROAD GAUGE RAILWAYS.	Miles
Completed	1844	Completed : Great Western, Bristol and Exeter, Cheltenham and Great Western (just completed), Bristol and Gloucester . . .	278
In progress	614	In progress	52
Projected	6918	Projected	1311
	9376		1641
Or, as . .	$5\frac{3}{4}$	To . . .	1

To strengthen the case of the narrow gauge, on this ground, it was advanced that already various lines in England and Scotland (like Mr. Braithwaite's *Eastern Counties'*) constructed with an intermediate gauge of 5 feet, had for the sake of uniformity been reduced to 4 feet $8\frac{1}{2}$ inches. Thus the proposition could never for an instant be entertained that to please an innovating minority, an overwhelming majority should alter arrangements which they had been at great cost to complete.

In the January of 1846, the Gauge Commissioners made a report, recommending,—

---

* 'Railways: The Gauge Railway Question,' &c. By Wyndham Harding, with a map. 4th ed.

(1) That the gauge of four feet eight inches and a half be declared by the legislature to be the gauge to be used in all public railways now under construction, or hereafter to be constructed in Great Britain.

(2) That unless by the consent of the legislature, it should not be permitted to the directors of any railway company to alter the gauge of such railway.

(3) That in order to complete the general chain of narrow gauge communication from the north of England to the southern coast, any suitable measure should be promoted to form a narrow gauge link from Oxford to Reading, and thence to Basingstoke, or by any shorter route connecting the proposed Rugby and Oxford line with the South Western Railway.

(4) That as any junction to be formed with a broad gauge line would involve a break of gauge, provided our first recommendations to be adopted, great commercial convenience would be obtained by reducing the gauge of the present broad gauge lines to the narrow gauge of four feet eight inches and a half; and we, therefore, think it desirable that some equitable means should be found of producing such entire uniformity of gauge, or of adopting such other course as would admit of the narrow gauge carriages passing, without interruption or danger, along the broad gauge line.

On the appearance of this judicious report the agitation of the two great parties whom it especially concerned, and of society at large, was indescribable. Articles and pamphlets of an acrimony unusual even in party warfare were published in every quarter; and the farces and extravaganzas of the theatres were full of allusions to the quarrel. Of the more eccentric literature of the contest, the 'Dialogues of the Gauges' first published in the 'Railway Record' may be read with amusement by the curious.

Beaten successively on all the engineering points, Brunel and his party endeavoured to persuade the public that their interests were concerned in maintaining a spirited

competition between broad and narrow lines.   This extraordinary view the engineer put before the Gauge
Commissioners in the following words:—

I think the spirit of emulation and competition (said Mr.
Brunel before the Gauge Commissioners) kept up between
different railway interests, both as regards the comfort and the
construction of the carriages, and the times and mode of
travelling, will do much more good to the public than that
uniformity of system which has been talked of for the last two
or three years.

After the publication of the Commissioners' Report, this
plausible fallacy was reiterated; and it was gravely maintained that the public would be benefited, if railway companies (composed of that same public) would lay down rival
roads side by side, and ruin each other by competition.
To this ridiculous proposition Mr. Thornton Hunt replied : *—

It is not possible.   There are not enough railways, nor
likely to be enough to create a real competition.   For the
most part railways branch off in different directions.   Where
they run in somewhat similar directions, the competition would
occur between very few parties.   Where parties are so few, so
well-organised, managed by councils possessing so much of a
deliberative character, and where the bad results of competition
would be shown so tangibly, and in such large amounts, a con-
t'nuous and injurious struggle for any length of time would be
practicably impossible.   'I look upon it,' says Mr. Laing, 'as
inevitable, that if a rival line is made, the two must sooner or
later agree to charge the rate of fares which will be the most
productive.'

This argument, based on the supposed advantages of
competition, is interesting, as it stands out to mark the

---

* 'Unity of the Iron Net Work :
showing how the last argument for
the break of gauge, competition, is
at variance with the true interests of
the public.'   By Thornton Hunt.
3rd ed. Smith, Elder, & Co., Cornhill.

extreme point to which Brunel was driven from the
ground on which, at the outset of the memorable war of
the gauges, he had taken his position.

Amongst laughable occurrences that enlivened the
committee rooms during the gauge contest, was a scene
occasioned by parliamentary counsel putting in as evidence,
before the committee on the Southampton and Manchester
Line, a printed picture of troubles consequent on a break
of gauge. The picture was a forcible sketch, that had
appeared a few days before in the pages of the 'Illustrated
London News.' Opposing counsel of course argued
against the production of the work of art as testimony
for the consideration of committee. After much argument
on both sides the chairman decided in favour of receiving
the illustration, which was forthwith put, amidst much
laughter, into the hands of a witness, who was asked if it
was a fair picture of the evils that arose from a break of
gauge. The witness replying in the affirmative, the engraving was then laid before the committee for inspection.*

Fortunately for the immediate peace and the permanent
interests of society, the conflict was concluded by legislation which in its chief principle accorded with Robert
Stephenson's views. By 9 & 10 Vict. cap. 57 (An Act for
Regulating the Gauge of Railways, August 18, 1846) it was
enacted :—

That after the passing of this act it shall not be lawful
(except as hereinafter excepted) to construct any railway for the
conveyance of passengers on any gauge other than four feet
eight inches and half an inch, in Great Britain, and five feet three
inches in Ireland. Provided always, that nothing hereinbefore

---

* Vide 'Railway Chronicle,' June 13, 1846.

contained shall be deemed to forbid the maintenance and repair of any railway constructed before the passing of this act on any gauge other than those hereinbefore specified, or to forbid the laying of new rails on the same gauge on which such railway is constructed, within the limits of duration authorised by the several acts under the authority of which such railways are severally constructed.

Possibly somewhat at the expense and to the detriment of the public, sections II. III. and V. of the same Act paid full measure of respect to existing broad gauge interests. There is ground for the opinion that too great consideration was displayed to the interests of individuals whose action threatened to be, and already had been prejudicial to the state. But it must be remembered that the gauge party was compact, united, and animated with a determination to fight to the last. Nor was its influence solely dependent on its spirit and organisation. In the houses of parliament it numbered many devoted and powerful adherents, and it had a strong hold on the opinions of those who, believing in the broad gauge as a system capable of supplying the public with greater and easier speeds than the narrow gauge, regarded the question from a selfish point of view, and placed personal comfort before commercial utility. Strong, therefore, within the walls of parliament, and strong without, the broad gauge party, even at the time of its overthrow, was to be conciliated. With 9 & 10 Vict. cap. 57, the gauge question, as far as the general public felt concern in it, was set at rest. An important professional question however was still to be discussed by engineers. A limit had been put to the construction of broad gauge lines. The question now to be considered was — how best to introduce the narrow gauge into broad gauge lines, so that the two systems might be worked together, where break of gauge could not be otherwise avoided? Ought two distinct pairs of lines for each gauge to be put down? or would it be better to use only three rails, one of them being common to both gauges? At first sight the choice, apart from the difference of original expense between putting down three lines or four lines, might seem unimportant. But practical men knew otherwise. Robert Stephenson's opinions on these questions were published in a report.

CHAPTER II.*

IRON BRIDGES.

Mr. Stephenson's large Practice in Iron Bridges—His Article on the Subject in the *Encyclopædia Britannica*—Modern Use of the Material—Early Bridges—First Iron Arch Bridges—Tom Paine's Bridge—Full Development of the Iron Arch Bridge—First Use of Wrought Iron—Suspension Bridges—Captain Samuel Brown—Mr. Telford—The Menai Bridge—Introduction of Railways—Consequent large Demand for Iron Bridges—Return to the Form of the simple Beam—Comparison of the three different Systems of Iron Bridges—Advantages of the Girder System—Cast-iron Girders—Compound Girders—The Dee Bridge—Royal Commission on Iron Railway Structures—Introduction of Wrought-iron Girders—Different Varieties of Girders—Examples—The Aire Bridge—The Benha Bridge—Last Work of Mr. Stephenson's Life, Restoration of Tom Paine's Bridge.

BRIDGES have always formed important works in the practice of the civil engineer, and there are scarcely any eminent members of the profession whose names are not associated with structures of this kind, of greater or less magnitude. Mr. Stephenson, in the course of his large railway practice, must have erected vast numbers, but his name is preeminently connected with three bridges, so important in their objects and so bold in their design, that they have acquired for their author a world-wide fame. These are the Britannia Bridge, carrying the Chester and Holyhead Railway over the Straits of Menai, in North

---

* This chapter is contributed by Professor Pole.

Wales; the High Level Bridge, for road and railway, across
the Tyne at Newcastle ; and the great Victoria Bridge over
the St. Lawrence in Canada. It is proposed to give a
brief account of each of these structures; but the description of their peculiarities will be simplified by
making some general preliminary remarks on bridge
building in iron—a material which now enters so largely
into the practice of the modern engineer. This will also
give the opportunity of referring to some other bridge
works of Mr. Stephenson's, which afford interesting and
instructive subjects for comment.

In this task it happens that we have an aid peculiarly
appropriate and useful. The subject is one in which Mr.
Stephenson took so great an interest, that he was induced,
in the midst of his heavy professional engagements, to
write the article ' Iron Bridges ' for the eighth edition of
the ' Encyclopædia Britannica.' The essay is, of course,
limited in its scope, and it is somewhat damaged by
typographical errors ; but still it treats the subject in its
historical, theoretical, and practical bearings in a very
clear and able manner, and with as much fullness as the
space at the author's disposal would permit. We are fortunate therefore in being enabled to base our account of
this subject on data of his own compiling, and frequently
to give his own words.

' The exclusive use of iron in the construction of bridges
is of modern date, though no other material is so peculiarly
adapted to such a purpose ; its use was, however, long
delayed, not so much because its advantages were not
appreciated, as from the great cost,and even impossibility,of
obtaining iron in large masses. It is now most extensively
employed in bridge construction, and though in elegance or

durability it cannot compete with stone, where the span is moderate, yet there are numberless cases where its adoption has been the means of solving many of the great problems of modern engineering. Its use has more especially become an absolute necessity in railway-bridge construction, where headway is so frequently of paramount importance, and where rapidity of execution is often a more necessary consideration than even economy or durability ; while the defective foundations that have so often to be contended with, render the lightness, the independent strength, and the pliable character of iron of the utmost value for such structures.'

The early attempts at forming a platform or bridge over an open space must have been by means of *beams*; that is, by pieces of timber, or some other material sufficiently long to stretch from one side to the other, and strong enough to carry the load required. But both the length and strength of such contrivances must have been naturally very limited ; and as the building arts advanced, a most important step was gained in the invention of the *arch*, by which a far greater span could be covered and a far greater weight supported, than by the simple beam. The arch is of very ancient date, having been traced back in Egyptian antiquities to many centuries before the Christian era. The Romans took advantage of it with great zeal and skill for the erection of bridges in vast numbers, and of large size and great constructive merit; and it has been used for the same purpose in all succeeding times down to the present day, wherever masonry has been the material employed for the structure. Yet it is a remarkable fact, that the development of the use of the

more modern material, *iron*, for the purpose of bridge
building, has just reversed this order of things. Iron
bridges were first made by imitating the more advanced
form of structure, the masonry arch, and for a long period
none but arched bridges were constructed in iron ; but
with the increase of knowledge in the use of the material,
and with the attainment of greater skill in its manufacture,
the design of the structures reverted to the primitive form
of the beam, in which by far the great majority of iron
bridges, including those of the most colossal dimensions,
are now made.

The history of iron bridges commences in the 16th
century, when such structures were first proposed in
some Italian works. In 1719 the subject was again
revived by Dr. Desaguliers, the well-known mechanical
philosopher, but nothing like an attempt at construction
was made till 1755, when an iron bridge was proposed
and partly manufactured at Lyons ; but the design was
subsequently abandoned from motives of economy, and a
timber bridge was substituted.

The credit of erecting the first iron bridge belongs to
this country, the work having been done immediately after
the time when, by the impulse given to the iron manufacture by smelting with coke, cast iron had superseded
timber in numerous details of mechanical construction.
This bridge, erected in 1779, was a cast-iron semicircular
arch of 100 feet span across the Severn at Coalbrookdale, a work which still stands, and which, considering
that the manipulation of the material was then completely
in its infancy, evinces a boldness and skill highly creditable to its designers.

Shortly after this, some propositions for an extension

of the principle were made by French engineers, but not one was carried into execution.

In 1794 two small iron bridges were erected in Germany, but the principle became much further developed in England before it was taken up in earnest in any other country. The celebrated Mr. Telford was one of the first to take advantage of the new material, and in 1796 he erected an iron bridge with a single arch of 130 feet span, also over the Severn, a little below Shrewsbury. In the same year, however, was finished a much larger iron bridge over the Wear near Sunderland, which Mr. Stephenson characterises as one of the boldest examples of arch construction in existence. It is also very remarkable in its paternity and history, its author being no other than the well-known Tom Paine, of sceptical and republican notoriety. This singular being, having a great aptitude for mechanics, proposed in 1790 to construct cast-iron arches, in what was then a novel manner, namely, of framed open panels in the form of voussoirs; and with characteristic energy he put his views to the test by making an experimental arch of 88½ feet span, which was exhibited at Paddington, and was completely successful. It happened that in the same year a committee was appointed for investigating the inconvenient and dangerous state of the ancient ferry in the middle of the harbour at Wearmouth; and as it was decided that a bridge should be built of cast iron, the ideas of Paine were adopted in its design, and part of the ironwork of his experimental arch was used in its construction. * The

---

* Mr. Murray, the engineer of the Sunderland Dock, who has had occasion to examine this bridge very carefully, believes that he has succeeded in identifying in it the particular portion of ironwork here referred to, which differs in manufacture from the rest.

span of this bridge, which is in one segmental arch, is
no less than 236 feet, only 4 feet less than the centre arch
of Southwark Bridge, the largest in existence, and yet
it contains only about one-fifth the weight of iron!

If (says Mr. Stephenson) we are to consider Paine as its
author, his daring in engineering certainly does full
justice to the fervour of his political career ; for, successful
as the result has undoubtedly proved, want of experience
and consequent ignorance of the risk could have alone
induced so bold an experiment ; and we are led rather to
wonder at, than to admire, a structure which, as regards
its proportions and its small quantity of material, will
probably remain unrivalled.

To complete the singular history of this bridge, it was
sold in 1816, by a lottery, for £30,000, £3000 more than
its original cost twenty years before; and certain alterations to it which will be described hereafter, formed the
last work of Robert Stephenson's engineering career.

In 1801 a remarkable design was given in by Messrs.
Telford and Douglas for replacing London Bridge by a
single cast-iron arch of 600 feet span; and the works
were even put in hand ; but the scheme was afterwards
abandoned on account of the great and inconvenient rise
that would be required in the approaches.

Iron bridges now began to be generally adopted. In
1802 a large arch of 180 feet span was erected over
the Thames at Staines, by the same engineers and on the
same plan as the Sunderland Bridge. The abutments were
of insufficient strength, and the bridge subsequently required supporting ; but it remained till the erection of the
handsome stone structure on a neighbouring site by Messrs.
Rennie in 1832, when it was taken down.

Iron bridges soon afterwards found their way into France. The Pont du Louvre, erected in 1803, was the first, and this was followed in succeeding years by the Pont d'Austerlitz and others, in Paris and in other parts of the country. In England they began to multiply fast. In 1816 Vauxhall Bridge was opened, and three years afterwards Southwark Bridge, which Mr. Stephenson declares stands confessedly unrivalled as an example of the cast-iron arch bridge, whether as regards its colossal proportions, its architectural effect, or the general simplicity and massive character of its details. The central arch is 240 feet span, the largest in the world; and the two side arches are each 210 feet. The bridge cost £800,000, and the quantity of iron in it is nearly 6000 tons.

At that time, therefore, the first form of iron bridge, that of the arch, may be considered as having arrived at its full development. Great numbers have been constructed in this and other countries, but there is nothing connected with them that needs further notice here.

Meanwhile, in the forty years that had elapsed since the first introduction of the iron bridge, great advancement had taken place in the manufacture of iron generally, and particularly in that of *wrought iron*. Cast iron had been eminently suitable for the arch bridge, as being specially adapted for resisting the strain to which arches were exposed, namely, direct compression or crushing. But to the use both of this form and of this material there was a necessary limit, from the massive construction and consequent great weight necessary, when employed for very large spans.

Wrought iron, by its greater tenacity and less liability to fracture, offered an extension of the limit of size possible for iron bridges. It was distinguished from cast iron by its property of withstanding a great *tensile* strain with a light weight of material, a property very valuable for bridge construction; the only problem being so to design the structure as to bring the material under this kind of strain. Hence arose a new construction of bridge, altogether differing from the arch, namely, the *suspension bridge*; the principle of which was, that the strain of the load was thrown upon a chain, keeping it in a constant state of tension, instead of upon an arch, in a constant state of compression.

The idea of forming a communication between opposite shores of a ravine or river, by suspending ropes across it, and attaching a roadway thereto, is of great antiquity, bridges of this description having existed in China, and indeed among less civilized nations, from time immemorial.

But the suspension bridge formed of iron chains, to the construction of which the resources of modern art and science have been applied, is of comparatively recent date. The first of which there is any account in England, or indeed in Europe, was a small foot bridge erected in 1741 across the Tees, near Middleton, in Durham, for the use of the miners. It was 70 feet long and 60 feet above the river, the roadway being 2 feet wide, of planking, with a handrail on each side; but no further particulars are recorded as to its details, and probably it was a very primitive affair.

In the commencement of the present century, however, the subject was taken up by an energetic naval officer,

Captain Samuel Brown, who appears to have had a knowledge of iron-work and of general construction worthy of an engineer of the present day. It was he who first introduced into the naval service the use of chain cables; and for this and his other services to the country, among which his part in the introduction and improvement of the suspension bridge was not the least important, he was knighted by the Queen in 1838.

Captain Brown's great improvement was the adoption of chains made of long iron bars instead of common link cables, which had been used up to that time. In 1813 he constructed a large model of a bridge on this plan, and in July 1817 he took out a patent for his invention.

The first bridge actually erected by him, and indeed the first suspension bridge ever constructed of much engineering pretensions, was the Union Bridge across the Tweed, five miles above Berwick, which was begun in August 1819 and opened in July 1820. It was a large bridge, being 450 feet span and having 30 feet deflection of chain. The roadway was 18 feet wide, and the chains were composed of link bars 15 feet long.

Captain Brown in following years also erected several other large structures on the same principle, among which, perhaps, the one best known is the chain pier at Brighton, opened in 1823.

In the meantime Mr. Telford was led to see the advantage of the suspension principle. About 1814 he had been requested to report on the practicability of forming a bridge at Runcorn over the Mersey. The plan of bar chains does not seem to have been then known to him, but he entered into a series of investigations and experiments to test the capabilities of the suspension principle generally, and its adaptability for the object he had in view. The strength of iron was not at that time so well determined as it is now, and one of Mr. Telford's objects was to ascertain, by direct experiment, the tensile power of the material when applied in the form of a suspended chain.

Nothing came of this proposition at the time, but Mr. Telford treasured up the knowledge he had gained, and a few years afterwards he had the opportunity of bringing it into practical application.

In 1819 he commenced the celebrated bridge over the Menai Straits, which magnificent work would, if he had done nothing else, have itself sufficed to render his name immortal. It was finished and opened in 1826, and its success made the principle popular; suspension bridges of large magnitude having since become very common.

About the year 1830, therefore, there were two great classes of iron bridges in use; the cast-iron arch bridge, heavy and of limited size, but rigid and strong; and the wrought-iron suspension bridge, lighter and much more capable of large expansion, but slender and less steady. Each kind had been brought to a state of great perfection, but the number of bridges built was not great, and the completion of a single large structure of the kind was an event in history.

But now arrived an epoch in civil engineering, which at once enlarged tenfold its sphere of action, and gave the application of iron for bridge purposes an entirely new direction. This was the introduction of railways.

Hitherto, bridges had been applied generally to common roads ; if the arch was adopted, inclined approaches were of small importance ; and in determining the rise of his arch, the engineer selected any headway he thought proper, while every other consideration was likewise made subsidiary to the problem of constructing the bridge itself. If the suspension bridge was chosen for the purpose, the passing load was light, and its speed of transit could be easily reduced to suit the comparatively unstable nature of the structure.

But on the introduction of railways, hundreds of roads, rivers, and valleys had at once to be spanned with bridges perfectly level, and of a strength and rigidity sufficient to allow the dashing across of the ponderous and swift locomotive, instead of the light coach or the quiet team. Moreover, a series of new conditions arose for these bridges, which complicated the problem still more.   Their time of construction was an important element ; so was economy of their first cost ; while every conceivable difficulty arose from their limited headway, their bad foundations, their oblique directions, their gigantic dimensions, and the necessity of bridging over navigable waters or crowded thoroughfares without interfering with the traffic upon them.   The number of bridges required also became something quite unprecedented ; Mr. Stephenson estimated that up to 1856 at least twenty-five thousand railway bridges must have been built in the United Kingdom alone.

The simple arch of masonry or brickwork was applied wherever it was practicable, but in many situations it was inapplicable, and the engineer, to whom the use of iron was now becoming every day more familiar, naturally turned to this material to supply the desideratum.

Of the two kinds of iron structures then in vogue, one, the suspension bridge, was, from its want of stability, quite out of the question ; but the other, the iron arch, was favourable in certain situations, where its well understood qualities of rigidity and strength warranted its adoption. Some of the most elegant and efficient railway bridges have been erected on this plan. The London and Birmingham and other early railways have several of this kind, and it is still used in situations where appearance is of importance, as the arch bridge may generally be made a handsomer structure than any other rigid form.

It was found, however, that the cast-iron arch bridge, from its great weight, and the small span of which it was capable within reasonable limits of cost, was but of comparatively limited application to railway requirements : hence it became necessary to discover some other kind of iron structure more generally suitable, and happily this was found by reverting to the earliest form of all, the primitive straight *beam*. This would seem, no doubt, a retrograde step from the elaborate and elegant structures on which so much scientific investigation and mechanical skill had been bestowed ; but the retrogression was only apparent, for no sooner had the beam been established as the normal model for railway bridges, than the attention of scientific and practical men was at once called to its development, and under this stimulus it soon outgrew its original simple form and dimensions. Improvements and extensions of the principle were gradually introduced, and the simple beam is now scarcely to be recognised as the parent of the many magnificent

structures which, far exceeding the largest arches in
dimensions, have become our most prominent monuments of engineering enterprise and skill. Still, however,
we cannot fail to be struck with the curious reverse
order of progress in the history of iron bridges, when we
find that the appliances resulting from ages of improvement have been rejected, to adopt a principle identical
with the earliest attempt of the uncivilized savage.

It may be well here to explain the principal points of
difference between the three different systems of bridges
above referred to, and to show on what grounds one of
them alone has proved so specially applicable to railway
purposes. When any structure is employed for carrying
weight over an open space, the laws of mechanics require
that the vertical forces due to the gravity of the load
should produce strains or thrusts in other directions more
nearly approaching to the horizontal. In an arch, the
essence of which is that it should curve downwards on
each side from the crown, a compressive strain is produced along the whole line of the curve, which, operating
at its extremities, tends to thrust the abutments outwards;
and this thrust must be efficiently resisted by massive
solidity and strength of the abutments, to keep the bridge
in equilibrium.

In a suspension bridge, this effect is reversed. The
essence of the structure is that the suspending chain must
curve upwards on each side from the centre, and must
sustain along its whole length a tensile strain, which tends
to draw the ends inwards; and this is usually provided
against by securing the ends of the chain firmly into the
ground on each side.

Now the beam is a sort of compound of these two principles. It is usually straight, neither turning downwards at the ends like the arch, nor upwards like the suspension chain; but it comprehends within itself the characters of both as regards the strains upon it ; for the effect of the load is to divide, in principle, the beam into two longitudinal parts throughout its whole length ; the upper part bearing a horizontal strain of compression, like the arch, and the lower a horizontal tensile strain, like the chain; these two strains being, moreover, as an essential condition of the equilibrium, equal and contrary to each other. Hence the strength of a beam is entirely self-contained, and all its horizontal forces are perfectly self-equilibrated ; the consequence of which is, that no resisting power whatever is required at the abutments or ends, further than is necessary to support the *vertical* pressure of the beam and its load, a condition capable of the simplest and easiest application. From these principles it will be seen that the advantages of the beam, or girder, as it is also called, for railway bridges consist in five great properties.

1. It supersedes the chain by its firmness and rigidity, being subject only to a slight deflection under its load, which is of no practical disadvantage.

2. Compared with the arch it has the great advantage of straightness, not requiring to be curved downwards at the ends, and so not only making a level road above, but also leaving a uniform height of headway underneath, which is often a vital necessity. It will be seen hereafter that it was this condition that determined the use of a beam for the Britannia Bridge.

3. As the beam requires no preparations in the

abutments for resisting any horizontal or oblique thrust, the construction of these parts of the bridge, particularly as regards their foundations, is rendered very much simpler, more expeditious, and less costly.

4. The ironwork required for a beam is generally very much less in weight than for an arch of the same span and strength.

5. The self-contained strength of the beam, and its capability of being fixed in many cases without scaffolding, much facilitate its erection; not only as saving cost and time, but also in avoiding interference with navigation or traffic below.

The earliest iron beams of which any account is preserved were used by Messrs. Boulton & Watt in building a cotton mill at Manchester in the year 1800; and as soon as confidence became established in the material, and the improvements in the manufacture of iron enabled large castings of this description to be made with tolerable certainty as to quality, and at reasonable price, iron beams soon began to supersede the use of timber for many building purposes, as being much less liable to decay, or to destruction by fire.

In 1822 Tredgold wrote his celebrated 'Practical Essay on the Strength of Cast-Iron,' the principal object of which was to define the theoretical laws that governed the construction of iron beams, and to put them into such a shape as should be useful to the practical mechanic and builder. Two or three years afterwards cast-iron girders of 50 feet span, the largest then constructed, were erected by Mr. Rastrick at the British Museum.

It was natural that the first engineer who had railways

of any importance to make, should first find out the applicability of the beam to railway-bridge construction; and accordingly the first bridges of this kind were erected by George Stephenson about the year 1830 on the Manchester and Liverpool Railway.

The girders used for this purpose were made entirely of cast-iron, in fact, were simple cast-iron beams, similar to those before used for other purposes; but as the object they had to serve soon became much more important, and the spans required much larger, more attention was called to the principles of their construction both from a theoretical and practical point of view.

The theoretical part was taken up by Mr. Eaton Hodgkinson soon after the erection of the first girder-bridges, and he corrected some errors that had been entertained as to beams of cast-iron, and established greatly improved rules for their proportions, by which their strength was much increased and their cost greatly reduced.

In a practical point of view the attention of engineers was soon drawn to the uncertainty and weakness of cast-iron, when exposed to a tensile strain in the lower flange of the girder. The proper function of cast-iron had been developed in the arch, namely, to withstand compression; for a strain in the contrary direction it was peculiarly unfitted, not only by its want of cohesive strength, but still more from the almost inevitable existence, in all large castings, of hidden flaws and defects. Little benefit was obtained by increase of thickness, for the treacherous character of the material increased rapidly with the mass in which it was cast; and the difficulty of uniting cast-iron rendered impracticable the attempt to build up such girders of separate castings, so that the new girder-bridges were limited in their dimensions to very moderate spans.

In order to meet this difficulty the girders were in some cases made double, so as to diminish the dangerous influence of possible unsoundness; but still an obvious necessity arose for some new combinations of the material which should meet the desired end with greater aptitude.

The first important contrivance springing out of the necessities of railway bridges was a modification of the cast-iron arch. The chief obstacle to the use of the ordinary arch was the practical difficulty of meeting the thrust at the abutments, and of obtaining the requisite stability in the foundations; a difficulty much enhanced by the diminution of rise or versed sine which the use of iron allowed, giving a consequent augmentation of the thrust, and a more unmanageable direction of its action. To meet this difficulty, in cases where headway was not of importance, the device was hit upon of connecting the two ends of the arch together by a wrought-iron tie rod, which, by taking upon itself a horizontal tension, deprived the ends of the arch of the tendency to thrust outwards, and so relieved the abutments of all except vertical pressure. The structure thus became essentially a girder, as it contained within itself the perfect equilibration of all its horizontal strains; and as the form resembled that of a bow, having the tie rod for a string, it was called the Bowstring Girder.

Another advantage followed from this construction. It was soon found that by suspending the tie rod strongly from the arch, it might be made to carry the rails at a lower level; the depth of the girder being thus above the roadway instead of below it; by which the attainment of one of the greatest and most troublesome requirements of railway bridges, namely headway underneath, was greatly facilitated.

The earliest railway bridge on this plan was designed by Mr. Robert Stephenson in 1834, and erected in 1835 or 1836, to carry the London and Birmingham Railway over the Grand Junction Canal near Weedon. This kind of structure has since been much used, and the finest example of it is the High Level Bridge at Newcastle-on-Tyne, of which a more complete account will be given hereafter.

But this construction was expensive and cumbrous; and attention became again turned towards the improvement of the simple cast-iron girder. The most prominent defect of this consisted, as already stated, in the weakness of the lower flange; and the most natural attempt to remedy the evil was by strengthening it with wrought-iron rods, so arranged as to take the tension upon themselves, and thus relieve the more defective cast metal from the tensile strain which it was so little able to bear. The wrought-iron rods were attached by screws at each extremity of the girder to its upper flange, and at the centre were brought down below the bottom flanges, being then tightened up by the screws to such a degree of tension as might be thought desirable. The girder was thus a compound one, of cast and wrought-iron together, and from the peculiar trussing up of the wrought-iron rods it was called the Trussed Girder. Such a beam, if made with due attention to the strains, was evidently less liable to accident than the simple casting, and was capable of application to much larger spans.

The first trussed compound girder, of 60 feet span, was erected about 1839 by Mr. Bidder, in conjunction with Mr. Stephenson, for carrying the Cambridge branch of the Great Eastern Railway (then called the Northern and Eastern) over the River Lea near Tottenham; others followed on the same and other lines, one of the best known being that over the Minories, on the Blackwall Railway. The plan was beginning to be somewhat extensively adopted in railway practice when an occurrence took place which at once checked its use, and which, from Mr. Stephenson's connection with it, must be noticed at some length. This was the memorable and fatal accident that occurred through the failure of the bridge at Chester, in May 1847.

The Chester and Holyhead Railway crosses the River Dee immediately after leaving Chester, and from the Chester station to a little beyond the crossing the line is also used, under an agreement, by the Chester and Shrewsbury Company, who, after running over this portion of railway, diverge to the westward by a line of their own.

The Dee Bridge, forming part of the works of the Holyhead line, was designed by Mr. Robert Stephenson, their engineer. The width of the river at this point is about 250 feet, and the railway is elevated nearly 40 feet above low water. The bridge was originally intended to consist of five brick arches, for which the piling was actually commenced; but apprehensions as to the foundations caused the engineer to change his design, and to substitute a bridge of iron girders, altering the number of

openings from five to three, and increasing their spans
accordingly. The bridge was considerably askew, forming
an angle of 51° with the river, or 39° with the perpen-
dicular crossing line ; and the length of the girders was
98 feet clear span. There were four main girders to each
span, twelve in all.

The girders were on the principle above described, i. e.
cast-iron trussed with wrought-iron tension bars. Each
was made in three lengths, bolted together, and was
3 feet 9 inches deep, or about one-twenty-sixth of the span,
having flanges at the top and bottom. The trussing or
tension rods, placed on each side, formed a chain of three
long links ; the middle link horizontal, and placed about
the level of the bottom of the cast-iron girder ; the two
outside links rising up obliquely towards their ends, which
stood at a height of about four feet above the top of the
girder, and were bolted to large shoulders or bosses, projecting upwards above its top edge. The lower parts
of these chains were caused, by means of screws, to
press upwards against the cast-iron girders, and so to afford
it support by suspension, the links forming essentially
suspension chains.

The girders were probably designed in 1845 or early in
1846. In September of the latter year one line of the
bridge was passable; on October 20 it was examined
and approved by the Government inspector, and immediately afterwards it was opened for traffic, not by the
Holyhead Company, to whom it belonged, but by the
Shrewsbury Company, who were the first ready to use
the bridge for public traffic. The Holyhead line was not
opened till some time after the accident. From the time
of the opening, the bridge was constantly used, not

only for Shrewsbury passengers, but also for heavy trains of materials for both lines ; but up to the day of the accident, May 24, 1847, nothing occurred to attract attention. It happened that about this time one of the Great Western bridges had been burnt down by cinders from an engine, and alarmed by this disaster, the authorities of the Chester Railway had laid down on the Dee Bridge about 18 tons of broken stone as a protection to its wooden platform. This was done on the afternoon of the day in question, and the first train that traversed the bridge afterwards was the fatal one. Leaving Chester, the engine passed safely over the first and centre openings, but when it arrived about the middle of the third opening, the left-hand or southern girder broke into three pieces, and the carriages fell into the river, at 36 feet below. Five people were killed, and all in the train more or less injured, except the driver ; the engine, which ran on beyond the fracture, being the only vehicle that remained on the line.

This accident made naturally a great sensation, not only from the gravity of the casualty, but from the importance of the consequences to railway engineering. It was felt that the bridge was upon its trial, and as it was soon found that nothing was defective in the *manufacture* of the ironwork or the quality of the material, the investigation became directed to the *principle* of the girder, and to the question whether the strength of beams of this description could be depended on.

The enquiry before the coroner was a very lengthened one. A great deal of engineering evidence was brought forward ; two referees, Mr. James Walker, civil engineer, and Captain Simmons, R.E., being also appointed by the Government to investigate the matter.

Mr. Stephenson was naturally looked to for his opinion, which he gave in a report addressed to the Railway Directors, and subsequently enlarged upon in oral evidence before the coroner. He stated that a few hours before the accident, on his way to Bangor, he had narrowly inspected every part of the bridge, and saw nothing to indicate weakness or imperfection. He confidently concluded that every part was firm and sufficient, a conclusion in which he conceived he was justified by the fact of the Chester and Shrewsbury traffic having been uninterruptedly carried on from October to May. He had examined carefully the appearances after the accident, and could arrive at no other conclusion than that the fracture of the girder arose, not from inability to support the weight, but from a violent blow given by the tender, which he conceived to have got off the rails, probably from the fracture of one of the wheels, while passing the bridge. Mr. Stephenson had full confidence in the proper strength of the bridge, in which he was confirmed by an extensive experience in the combination and use of similar structures, tried under circumstances that demonstrated their capabilities to meet all the ordinary contingencies of railway traffic. An objection had been made that the wrought-iron tension rods did not act well in concert with the rigid cast-iron girder; but he had well considered this, and had had experiments made which had satisfied him there was no force in the objection. If the tension rods were properly screwed up, they would bear the whole strain from the weight passing over the bridge, and would thus take the place of the cast-iron girder, and that was what he sought most to rely upon. He did not maintain that the two principles could be brought into strict union at one and the same time, but he urged that they might mutually aid each other. Mr. Stephenson added that he had erected, in twenty years, more iron bridges than any other member of the profession, being more partial to them, and this was the first failure he had had, large or small.

Mr. Locke and Mr. Vignoles supported the opinion of Mr. Stephenson that the fracture arose by a blow, and that the girder was sufficiently strong. Mr. Locke did not, however, like iron bridges, preferring those of brick or stone.

Mr. Robertson, the engineer of the Chester and Shrewsbury line, reported to his Directors his conclusion that the girder broke in the middle from its weakness to resist the strain, increased by the laying on of the extra ballast immediately before.

The referees appointed by Government made their report on June 15, 1847. After stating the facts and describing the bridge, they considered the strain of the girder and the action of its parts, the effect of temperature, of oscillation, &c., and summed up their opinion, that though the bridge was of sufficient strength if the cast and wrought-iron were supposed to act together, each taking its equal proportion of the strain, yet neither, separately, was sufficient for perfect stability; and that there was great difficulty in ensuring the joint action. They did not agree in Mr. Stephenson's view that the fracture was caused by a blow.

This report was communicated to the coroner's jury the last day of their sitting, and seems to have guided them in their decision. They gave a verdict, through Mr. E. Walker, their foreman, of accidental death; adding, however, that they were of opinion the girder broke from being made of a strength insufficient to bear the pressure of quick trains passing over it; that they considered the remainder of the bridge unsafe; and that for the security of the public, they recommended a Government enquiry as to the safety of such bridges in general.

The propriety of this verdict was questioned at the time, but it must be recollected that at that period the nature of the strains in compound girders was very little understood, and therefore we may be quite prepared to admit that the girder may have been imperfect in design without in the least disparaging Mr. Stephenson's credit as an engineer.

Indeed, we cannot offer a better description of the defects of this kind of girder than is given by Mr. Stephenson himself in his Essay of 1859. He says:—

The determination of the strength of such girders is a difficult task. They are, in fact, compound girders, formed by combining the truss with the simple girder, the upper flange doing duty as a compression bar in both systems, and being thus subjected to two independent strains. It is evident, therefore, that if the upper flange is simply proportioned to its duty, as the top flange of the simple girder, it will be of insufficient strength for its additional duties as part of the truss. It has been argued that from the perfect union of the top flange with the vertical rib, a considerable portion of the whole girder might be taken as forming part of the truss. It is, however, evidently impossible by calculation to say how far such assistance may be relied on; and a still greater objection exists in the fact that such girders consist of two systems, the ultimate deflections of which are utterly different; the girder, for instance, may be broken before the truss attains half its ultimate deflection or has done half its duty. The objection to this girder is common to all girders in which two independent systems are attempted to be blended; and, as a general principle, all such arrangements should be avoided.

It is useless (adds Mr. Stephenson) to say more on the subject of this form of girder, as since the adoption of wrought-iron for girders they have been entirely superseded; they were designed when no other means existed of obtaining iron girders of great span, and the melancholy accident which occurred at Chester is the only existing instance of their failure.

Mr. Stephenson, in his evidence before the Iron Railway Structure Commission,* further explained the objection to the design of these girders, which, in the more advanced state of our present knowledge, is clearly perceived to be the want of a due provision for withstanding the inward thrust of the ends of the wrought-iron ties. The bolts to which the tie bars (which acted, in fact, as suspension chains) were attached, were elevated some four feet above the general level of the top of the cast-iron girder, and no direct solid member for resisting the compressive strain existed between them. The cast-iron girder itself, being of such a small depth in proportion to the length, was very weak, and, as Mr. Stephenson stated, the principal reliance was on the wrought-iron bars; but when the heavy strain came upon these, tending, as in a suspension bridge, to draw their ends inwards, nothing existed sufficiently strong to keep them apart, and consequently acting with a strong leverage and in a most trying manner upon the top flange of the girder, they compressed it beyond its strength, and broke it through. That this was the true explanation of the failure is now clear from the form of the two fractures, which (although this does not seem to have been noticed at the time) are identically of the description peculiar to the case where the

---

* Questions 832, 881, 894.

upper flange of a beam is broken by a compressive
strain beyond its resisting power.*

Mr. Stephenson on discovering this defect at once took
measures to provide against it in other girders on this
plan, by adding properly shaped compression pieces of
cast-iron to the top of every girder, so as to fill in, solidly
and strongly, the space formerly open between the ends
of the ties; and the bridges thus strengthened have never
shown any signs of failure.

The Dee Bridge was altered by having inclined struts,
bearing against the masonry, placed under each girder,
so as to afford support in the middle; and other bridges,
made about the same time, were strengthened in like
manner.

The last large girders on this principle were some of
96 feet span, made under Mr. Stephenson's directions, in
1847, for the Florence and Leghorn Railway, crossing the
Arno, and in these the proper improvements were introduced in the original design.

The matter, however, did not stop here. The Government Commissioners of Railways, on receiving the report
of the two engineers to whom they referred the investigation of the accident, became alarmed about the iron
bridges used on railways generally. On June 23, 1847,
they addressed a circular to the secretaries of the different companies, requesting a return to be made of the
iron bridges on all lines then working or constructing,
giving their dimensions and particulars of their construc-

---

* Compare the sketches in the   in Hodgkinson's edition of 'Tred-
Report of the Commissioners of   gold on Cast Iron,' 1846, p. 429.
Railways, 1848, p. 110, with that

tion; and expressing a hope that wherever the security of such structures was at all doubtful, the companies would take measures to add to their stability, and in the meantime would direct the speed of the trains to be reduced in passing. A few days afterwards they published a minute, to the effect that they repudiated all responsibility for the strength of iron structures which had been inspected and passed by their officers, inasmuch as these gentlemen had only the opportunity of a superficial observation, and no sufficient control over the design.

Another matter of public interest followed. The Railway Commissioners, acting on the suggestion of the coroner's jury, passed a minute calling the attention of the Government to the uncertainty which existed respecting the conditions to be complied with in employing iron in engineering works, and in particular to bridges which had to be traversed by loads of extraordinary weight with great velocity. They had reason to believe, they said, that much difference of opinion existed among the most eminent engineers of the time as to the proper form and dimensions to be given to railway girders of iron for bridge purposes; and they considered it desirable that the subject should be thoroughly investigated by a Commission, to be composed of scientific men and practical engineers, who should be appointed by Government, and should be requested to arrive at such principles, and to form such rules, as might enable the engineer and the mechanic to apply the metal with confidence in their respective spheres.

The Commission was appointed by Royal Warrant on August 27, 1847, and consisted of Lord Wrottesley, Professor Willis, Captain James, R.E., Mr. George Rennie,

Mr. (afterwards Sir) William Cubitt, and Mr. Eaton Hodgkinson, with Captain Douglas Galton, R. E., for secretary. They spent nearly a year in examining witnesses, making theoretical investigations, trying experiments, and collecting a great mass of information on the subject, which was afterwards published in a Blue Book of 435 pages, accompanied by a large collection of lithographed plans.

Mr. Stephenson was one of the principal witnesses. He gave much information as to the nature and properties of iron—the construction of girder-bridges, particularly those of the kind used over the Dee—the effect on them of passing trains, &c. &c.; but he strongly impressed upon the Commissioners that any attempt to introduce restrictive legislative enactments in regard to the use of iron in railway structures would be highly inexpedient, and would act prejudicially on professional enterprise and skill. 'My opinion,' said he, 'is rather strong that a collection of *facts* of all kinds is highly desirable, in reference to the shape of girders; but I am convinced that the Commissioners will have infinite difficulty in laying down anything like rules. I cannot conceive myself being tied down in executing such a line, for instance, as the Holyhead, or the London and Birmingham. I cannot conceive myself going on successfully, and being tied down by preconceived rules, or limitations as to the extent to which cast-iron should be used, and the forms that it should be used in. I think a collection of facts and observations would be most valuable; but if you attempt to draw conclusions from those facts, and confine engineers, even in a limited way, to those conclusions, I am quite sure that it will tend to hamper the profession very much.'

In addition to his oral evidence, Mr. Stephenson furnished a valuable statement of the experiments on iron undertaken at his direction for the High Level Bridge at Newcastle; and also brought up before the Commission two of his chief assistants, namely, Mr. Edwin Clark, who gave a full account of the Britannia and Conway Bridges; and Mr. Charles Heard Wild, who described other large girders made under Mr. Stephenson's direction. Mr. Fairbairn and Mr. Hodgkinson also gave full accounts of the comprehensive experiments conducted for the great tubular bridges, so that Mr. Stephenson's opinions and works may be fairly said to have formed the largest and most important part of the information collected by the Commissioners.

Many other witnesses were examined, skilled in the engineering of ironwork, among whom were Mr. Brunel, Mr. (now Sir) Charles Fox, Mr. Locke, Mr. Rastrick, and Mr. Charles May; and much information was also collected in the form of written statements. Professor Willis, aided by Professor Stokes of Cambridge, contributed an elaborate theoretical paper on the deflection of beams under moving loads; and a comprehensive series of experiments was tried by the Commission on various points coming within the scope of their enquiry.

The Report of the Commissioners was presented to Her Majesty at the end of July 1848. They considered that bridges should be made somewhat stronger to meet the additional strains from moving loads; and they recommended that the greatest load should in no case exceed one-sixth of the stationary weight which would break the beam when laid on the centre. They also pointed out that weight is an advantage in enabling a structure to resist concussions. As to designs of iron railway bridges, they merely stated the facts and opinions laid before them by engineers. They testified to the careful and scientific manner in which the forms and proportions of the great tubes of the Conway and Britannia Bridges had been elaborated. They thought that wrought-iron plate girders generally appeared to possess and to promise many advantages. They found engineers to be for the most part favourably disposed towards them; but as no experience had yet been acquired of their powers to resist the various actions of sudden changes of temperature, vibrations, and other causes of deterioration, they were unable to express any opinion upon them. With regard to trussed cast-iron bridges, like that of the Dee, they found that difficulties arose from the different expansions and elongations of the two metals, and considered that the greatest skill and caution were necessary to ensure the safe employment of such combinations. They also stated that there existed a great want of uniformity in practice in many most important matters relating to railway engineering, which showed how imperfect and deficient it yet was in its leading principles (a reproach which unfortunately is almost as applicable in 1864 as it was in 1848); but considering that the attention of engineers had been sufficiently awakened to the necessity of providing a superabundant strength in railway structures, and also considering the great importance of leaving the genius of scientific men unfettered for the development of a subject so novel and so rapidly progressive as the construction of railways, they concurred in Mr. Stephenson's opinion that any legislative enactments with respect to the forms and proportions of the iron structures employed therein would be highly inexpedient.

After the completion of the London and Birmingham Railway, notwithstanding the progress of iron roads in all directions, no important step seems to have been made in the improvement of iron bridges, until the epoch of the Britannia Bridge, the erection of which initiated a complete revolution in this branch of engineering science. As an account of this great structure will be given hereafter, it is only necessary to notice here the effect which it had upon bridge construction in general.

About 1845, when the experimental investigations commenced, the only forms of iron bridges used for railway purposes were the cast-iron arch, the simple cast-iron girder, the trussed compound girder, and the bowstring girder. In all these cast-iron had been the principal element, very little attention having been paid to wrought-iron as a material for girders, although its use had become common and was well understood for suspension chains.

Wrought-iron had indeed been used by Smeaton, in conjunction with wood, to form beams, by bolting an iron plate between two half balks of timber. The plate, being set vertically, contributed important strength to carry the load, while the wood furnished the necessary lateral stiffness. This kind of beam was, from its peculiar construction, called the 'flitch' or 'sandwich' girder, and it was used subsequently—in about 1839 or 40—in forming beams of thirty or forty feet span on the Cambridge branch of the Eastern Counties Railway. Wrought-iron beams, of analogous shape to those of cast-iron, had also been constructed for iron ships, and other purposes, and Mr. Stephenson had himself used them, about 1841, in a small bridge on the above-mentioned Cambridge line; but these were probably the only instances in existence of the use of wrought-iron in railway bridges. Very little was known as to the proper application of the material. The principles of its strength when applied to girders were quite undetermined; no such thing as a *constructed* beam of any scientific pretensions or any large span had been imagined; and even the process of connecting wrought-iron plates together by riveting was scarcely known beyond the boiler and iron shipping trade.

The experimental investigations, however, which were conducted for the Britannia and Conway Bridges threw quite a new light on the subject. From the time of the abandonment of the large arch which Mr. Stephenson at first proposed, it had become evident that cast-iron could not be applied, and that wrought-iron was the only material from which any real success could be expected; and it was therefore to the investigation of the properties of this material, and the best manner of using it, that the experimental enquiries were directed. They had the effect of thoroughly developing the powers of wrought-iron, of making known its peculiar properties, of rendering its use perfectly amenable to theoretical calculation, and of proving in the most conclusive manner its special applicability to girders for bridges of any magnitude. And they further showed the practicability of building up or constructing, in that material, girders of almost any strength and size, by only exercising a skilful and careful attention to the details of the design, based on a correct scientific knowledge of the nature and distribution of the mechanical forces acting throughout the structure. And it is worthy of remark how thorough and how perfect these investigations were; for although nothing was known of wrought-iron girders before they were undertaken, and although since their date wrought-iron girders have come into very general use under the greatest variety of forms, and in preference to all other systems of iron-bridge construction, yet nothing essentially new or important has been added to our knowledge of the principles of their construction beyond what was developed in these investigations, the records of which comprise indeed almost the whole useful information, theoretical or practical, we possess on the subject.

To this date, then, may be referred the first use of *wrought-iron girders* for bridge construction—the greatest step made in iron bridges since their original introduction; as giving them not only the capability of application to larger spans, but making them cheaper, more secure, lighter, more convenient of erection, and clearer in the headway; advantages almost incalculable for railway purposes.

The experiments were commenced in the middle of 1845, and early in the next year so much progress had been made as to lead to the proposition of hollow plate iron girders for railway bridges of considerable span. In July 1846 Mr. Stephenson gave instructions for a bridge on this principle, but with a cast-iron top, for a road at Chalk Farm, crossing over the North Western Railway. This bridge was sixty feet span; it was completed in March 1847, and was the first actual application of hollow wrought-iron girders to the construction of bridges.

Meantime Mr. Fairbairn, who had previously made drawings for a bridge of this kind, foreseeing that the use of hollow plate girders would be considerably extended, proposed to Mr. Stephenson to take out a patent for their application. To this Mr. Stephenson consented; but being averse to his own name appearing in the patent, he refused to accept any share of the profit, though he consented to pay half the expenses of obtaining the patent.* It was taken out October 8, 1846.

Mr. Fairbairn soon began to put the plan into operation, and in July 1847 completed bridges on this principle at Blackburn and Bolton in Lancashire, which answered well. The further progress of the designs for the large Britannia and Conway tubes, and their ultimate success, gave to the engineering world more complete confidence in the use of wrought-iron for bridge girders, and a few years more made its application universal.

It would be foreign to the object of this work to follow out in detail the wide and rapid progress of the art of iron bridge building after the epoch we have been considering; it will suffice to give a general view of the state it has now attained, and to describe briefly some of the numerous varieties which have sprung up in the construction.

Iron bridges may now be divided into three great

---

* Clark's Britannia Bridge, p. 812.

classes — namely, Iron Arch bridges, Suspension bridges, and Iron Girder bridges.

The first two of these have already been sufficiently described, and we may therefore confine our attention to the third or Girder class, which is a very large one, and may be subdivided into several species somewhat as follows:—

    1. Solid beams.
    2. Trussed cast-iron girders.
    3. Bowstring girders.
    4. Simple I-shaped girders.
    5. Tubular or hollow plate girders.
    6. Triangular framed girders.
    7. Lattice girders.
    8. Rigid suspension girders.

A beam or girder is distinguished from other means of bridging space by containing within itself the double horizontal strains, the top part of the beam being under compression and the lower part under tension. Hence every beam may be considered as consisting of three distinct parts, each of which has its own special office to perform; first the *top member*, which has to resist crushing; secondly, the *bottom member*, which has to resist being torn asunder; and thirdly, the *vertical part*, which has to connect these two together, and to combine the beam into one structural whole.

In the Solid Beam no distinction is made between the functions of the three parts, the top and bottom merging insensibly into the vertical connecting part. This kind of beam is represented by a stone lintel or a wooden floor joist, the form being never used in iron, on account of its wasteful distribution of material.

The Trussed Cast-iron Girder, which has already been sufficiently described, is no longer in use.

Of the Bowstring Girder, with the bow in cast-iron, we have also given an account; but we may add here, that as soon as wrought-iron came into use, large bowstring bridges were made entirely of this material. Among these may be instanced one of 200 feet span, built by Mr. Brunel in 1849, to carry a branch of the Great Western Railway over the Thames near Windsor; and two by Messrs. Fox and Henderson, carrying the North London Railway over the Commercial Road and the Regent's Canal at Stepney, about 165 feet span, built in 1848.

The I-shaped Girder was one of the earliest used in cast-iron. In it, economy of construction is aimed at by accumulating metal in the shape of flanges at the top and bottom, and connecting them by a vertical rib in the middle of their width, so as to give the whole section the shape of the letter I. This form was also early imitated in wrought-iron beams, and is still one of the simplest and best that can be used when the span is small.

As soon, however, as very moderate limits are exceeded, difficulties arise of a practical nature, in consequence of the distortion of form to which the simple I-shaped girder would be liable, if of great length and supported only on two distant and limited bearing surfaces, more especially as expansion and contraction prevent any rigid attachment even on these.

The Tubular or hollow construction of Wrought-iron Plate Girder was a step in advance to meet this difficulty. It differs from the I-shaped Girder only in that the top and bottom members are connected by two vertical plate ribs, one on each side, instead of a single one in the middle, so that the whole forms a sort of hollow tube, which is a stiffer and otherwise superior construction for large spans. Indeed, this plan may be carried to so large a size that, as in the Britannia and Conway Bridges, the engine and train may pass along inside the tube.

In the Triangular-framed Girder, the vertical rib, connecting the top and bottom members, instead of being composed of plates, is formed of a series of frames of a triangular shape, fastened to the top and bottom with large bolts. This form was first tried about 1850, at the London Bridge Station of the South Eastern Railway, and has since been a great deal adopted, with much success.

The largest bridge on this plan is one of 240 feet span, erected in 1852 on the Great Northern Railway over a branch of the Trent near Newark. A structure also very remarkable is a viaduct at Crumlin in Monmouthshire, erected in 1857. It consists of a series of seven triangular-framed girders, each 150 feet span, crossing a wide valley at an altitude of 200 feet, and supported by piers of framed ironwork, of great lightness of construction. The singular appearance of this structure can scarcely be imagined without seeing it, and it is certainly one of the engineering curiosities of Great Britain.

In the Lattice Girder, the vertical plates are replaced by a number of bars crossing each other so as to form a lattice-work, the strength of these bars being proportioned, by known rules, according to their places in the girder. The lattice principle was early used to a great extent for timber bridges, particularly in America. One of the largest, as also one of the earliest structures on this principle in iron, was the bridge or viaduct erected in 1855 on the line of the Dublin and Belfast Railway over the river Boyne near Drogheda. It consists of three spans, the centre 264 feet, and the sides 139 feet each, the height above high water being 90 feet.

The Rigid Suspension Girder is placed in a separate class on account of its use by Mr. Brunel in two magnificent iron bridges of gigantic dimensions and economic construction over the Wye at Chepstow, and the Tamar at Saltash. The former, completed in 1853, has a single span of 300 feet, at a height of 46 feet above high water. The girder is, in fact, a rigid suspension bridge, the tension of the chains being resisted, not in the usual way, by anchorage in the ground at each end, but by a huge cylindrical wrought-iron strut or column, stretching across from side to side, at a height of 50 feet at the centre, above the roadway. The whole is well braced together, and it thus forms a colossal trussed girder.

The Saltash Bridge, with the viaduct which forms its approach, carries the Cornwall Railway over the estuary and valley of the Tamar near Plymouth. The whole structure comprises nineteen openings, and is 2,240 feet long; but the bridge itself, crossing the river, consists of two spans of 450 feet each, at a height of 100 feet above the water. The main girders, or trusses, are in principle analogous to those at Chepstow; but here the compression tubes, which resist the pull of the suspension chains, are curved instead of being straight, the rise of the tube being equal to the drop of the chains. The tube thus partakes of the nature of an arch, and in fact the whole girder is a kind of intermediate between the Chepstow truss and the old bowstring girder.

The centre pier was a work of considerable difficulty on account of the great depth of water. The substructure is a solid cylindrical pillar of granite, 35 feet in diameter, resting on a rock foundation 86 feet below high-water mark. It was built in a coffer dam or cylinder of plate iron sunk through the bed of the river till it rested on the rock below, and then emptied and kept clear of water partly by pumping and partly by compressed air, to allow of the construction of the granite column inside.

The bridge was commenced in 1853, and was opened in May 1859 by H.R.H. the late Prince Consort, by whose permission it was called the Royal Albert Bridge.

Subsequently to the erection of the great works in North Wales, Mr. Stephenson designed three other iron bridges of considerable magnitude. The first was over the Aire, on the York and North Midland Railway, erected in 1850. The span was 225 feet, and Mr. Stephenson adopted the tubular girder, similar to the Welsh bridges, the trains passing inside the tube; but in this bridge, the span being so much smaller, the cells were dispensed with, and the top and the bottom were formed of simple plates. The tubes were originally constructed of a tapering section, narrower at the top than the bottom, but after they were erected and the bridge was opened, the narrowness at the upper part was objected to by the Government Inspector, and the top plate had to be cut through longitudinally and widened *in situ* by the insertion of a strip of iron all along; a delicate and unprecedented operation, but which, under great care, was perfectly successful.

In 1855 Mr. Stephenson erected a large wrought-iron bridge on the Egyptian Railway over the Damietta branch of the Nile near Benha. It consists of ten spans or openings, each of 80 feet, except the two centre ones, which are 60 feet each, and are made to open, forming what is called a swing bridge, one of the largest hitherto attempted. The girders for this bridge were also tubular, but from their small size the roadway is carried upon the top of the tubes, and not in their interior. The total length of the swing beam is 157 feet ; it is balanced at the middle of its length on a large central pier, so that when open to the navigation, a clear water-way of 60 feet is left on either side. Each half of the beam sustains its own weight as a cantilever 66 feet long.

The piers consist of wrought-iron cylinders, 7 feet in diameter below the level of low Nile, and 5 feet diameter above that level. They were sunk by the pneumatic process to a depth of 33 feet below the bed of the river, through soil of a peculiarly shifting character, and were filled in with concrete. There are six of these cylinders in the central pier which supports the swing bridge, and the adjacent piers on either side of the centre have each four cylinders. Each of the remaining piers has two cylinders only. This plan of obtaining the foundation of piers by sinking large iron cylinders has been a most important modern advance in bridge-building. One of the latest examples is in the bridge carrying the Charing Cross Railway across the Thames.

The beams or tubes are 6 feet 6 inches deep, and 6 feet 6 inches wide at the bottom, tapering to 6 feet wide at the top. They rest at their ends on rollers working between planed surfaces, to admit of the motion caused by expansion and contraction. The tubes carry a single line of railway on their tops, the rails being laid on longitudinal sleepers; and there is also a roadway four feet wide on either side, supported by wrought-iron brackets bolted to the sides of the tube.

The revolving machinery for the swing part of the bridge consists of a turntable 19 feet diameter, running upon eighteen conical rollers, connected by what is called a 'live ring.' The whole of this machinery is most carefully fitted and susceptible of the most accurate adjustment. The bridge is turned by a capstan connected by gearing with the moving parts, and which can be worked with facility by two men.

In 1853 Mr. Stephenson took up the subject of the Great Victoria Bridge over the St. Lawrence in Canada, a work of immense magnitude, of which a separate notice is given in a subsequent chapter.

The last work of his life had to do with one of the earliest iron bridges ever erected, and one in which he had always taken particular interest. This was the strengthening, and indeed almost the entire reconstruction, of the celebrated bridge over the Wear at Sunderland, which has been already noticed in the early part of this chapter.

At the request of the authorities of the town Mr. Stephenson had several times examined this bridge, and expressed his conviction that its stability was extremely precarious; and as they concurred in this opinion,

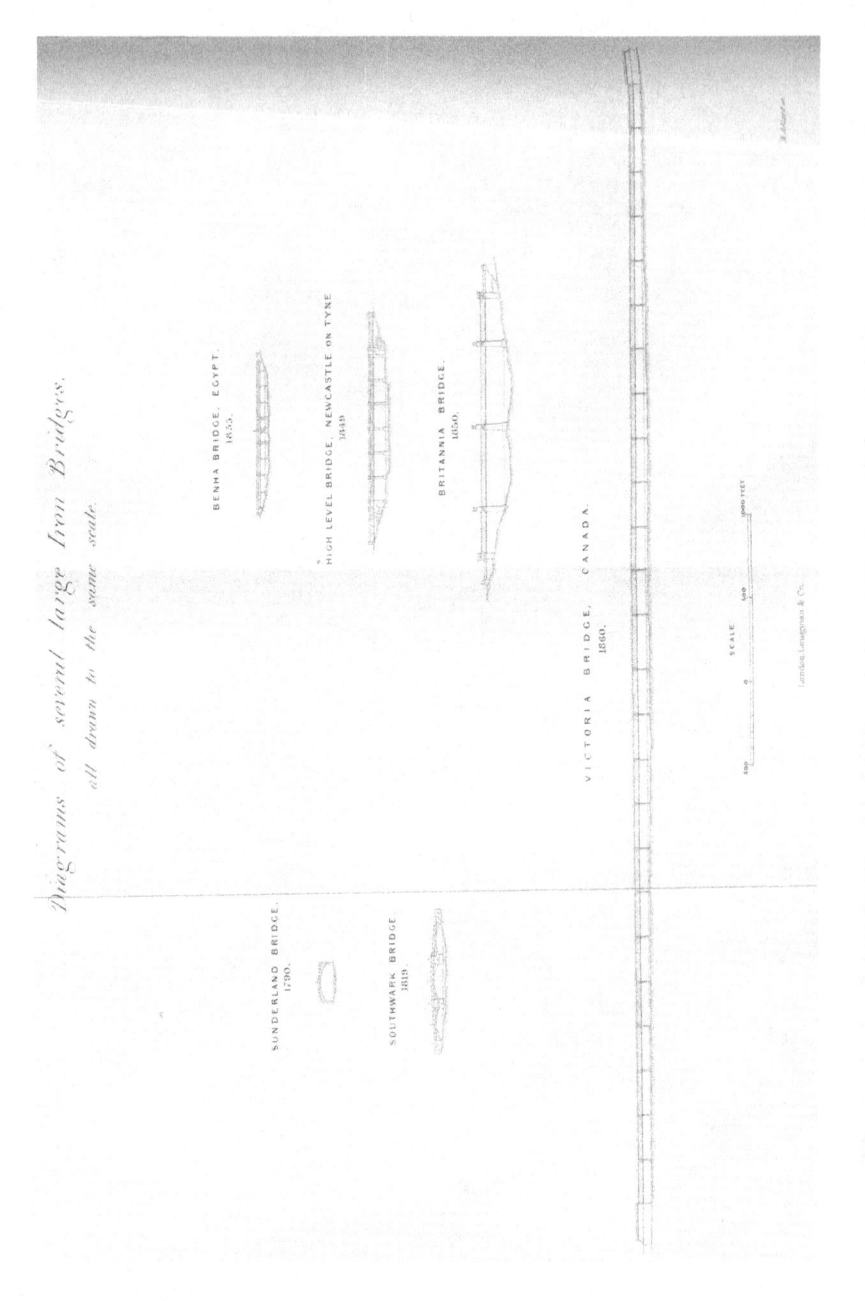

Diagrams of several large Iron Bridges,
all drawn to the same scale.

SUNDERLAND BRIDGE. 1793.

SOUTHWARK BRIDGE. 1819.

BENHA BRIDGE, EGYPT. 1855.

HIGH LEVEL BRIDGE, NEWCASTLE ON TYNE. 1849.

BRITANNIA BRIDGE. 1850.

VICTORIA BRIDGE, CANADA. 1860.

SCALE

London, Longman & Co.

he was in the year 1857 requested to undertake its repair. He was then in failing health, and reluctant to burden himself with further work, and he accordingly entrusted the details of the operation to one of his earliest friends and assistants, Mr. G. H. Phipps, by whom it was carried to completion; Mr. Stephenson, however, giving his advice and opinion, and occasionally visiting the bridge during the progress of the works. The work consisted of the introduction of three new tubular arched ribs of wrought-iron between the original cast-iron ribs of the bridge; the latter being firmly bolted to the new girders, and thus being relieved of the chief part of their load.

The width of the bridge between the hand-railings was also increased from 32 feet to 41 feet 1½ inch, and the road over the bridge and on the approaches was much improved, its inclination, or gradient, being lessened from 1 in 17 to 1 in 50.

One of the chief difficulties was the construction of a timber scaffolding, supported by pile work from the bed of the river, which should be sufficiently substantial to carry the weight of the arches, and should at the same time allow of the passage, with their masts standing, of the large amount of shipping frequenting the port; this scaffolding also formed a temporary bridge, both for carriage and foot traffic, during the progress of the alterations.

The work was let by contract, towards the end of 1857, to Mr. B. C. Lawton, by whom it was satisfactorily completed, and the bridge was re-opened for traffic in the summer of 1859, a few months prior to Mr. Stephenson's death.

The accompanying plate contains diagrams of several large iron bridges all drawn to the same scale, from which a comparison of their respective magnitudes can be made. The two small figures are the Sunderland Bridge and the Southwark Bridge; the others represent the four principal bridges erected by Mr. Stephenson, namely the Benha Bridge, the High Level Bridge, the Britannia Bridge, and the Great Victoria Bridge in Canada.

                                                    W. P.

Drawn by S. Russell.

Britannia Bridge, North Wales.

Etched by J. Adlard.

CHAPTER III.

THE BRITANNIA BRIDGE.*

The Port of Holyhead — The Holyhead Trunk Road — Interruption by the Menai Strait — Attempts to establish a Passage — Telford's Suspension Bridge — Introduction of Railways — Chester and Holyhead Railway — Proposal to use Telford's Bridge for Railway Purposes — Mr. Stephenson designs an Independent Bridge — The Britannia Rock — Proposal for a Bridge of two Arches — Opposition in Parliament — First Idea of the Tubular Construction — Its Novelty — Preliminary Experiments : Mr. Fairbairn and Mr. Hodgkinson — Important Principles derived from the Experiments — Mr. Stephenson's Report — Commencement of the Masonry — Further experimental Inquiries — Means of placing the Tubes in their Positions — Contracts for the Tubes — Their Manufacture — Floating and Raising — Description of the Bridge — Principle of Continuity — Tubes — Mr. Stephenson's Explanations of Peculiarities in their Construction — Towers and Abutments — Architectural Design — Cost — The Conway Bridge.

THIS celebrated structure has for its object to carry the line of the Chester and Holyhead Railway, the main artery of communication between the English and Irish capitals, across the Straits of Menai, which separate the island of Anglesea from the mainland of North Wales.†

---

* This chapter is contributed by Professor Pole.

† After the completion of the bridge, a full account of it was published by Mr. Edwin Clark, the resident engineer, with the sanction and under the supervision of Mr. Stephenson himself, the object being 'to preserve the history of a conception as remarkable for its originality as for the bold and gigantic character of its application.' Mr. Stephenson gave, in his own words, an account of the early history of the design, and he paid the author of these chapters the compliment of

The port of Holyhead, lying at the western extremity of the island, and forming the nearest point of land to Kingstown Harbour, has always been considered the most eligible place of departure for the passage across the Irish Channel, when certainty and speed of transit have been concerned. Liverpool has, it is true, carried on hitherto, and will doubtless continue to carry on, a large trade with the Irish capital by direct steamers, but it appears certain that in this case, as in the traffic with the Continent, that route must always be considered of the most importance which involves the least exposure to the perils and comparatively slow navigation of the sea.

Long before railways were thought of, the great Holyhead Trunk Road had made the fame of the engineer who constructed it, Thomas Telford; and this work presented, in one or two points of its course, difficulties so analogous to some of those which were vanquished in the Britannia Bridge, that they may be treated as common to the works of both engineers.

The island of Anglesea is separated from the mainland of Carnarvonshire by a narrow Strait, deeply sunk below the general level of the land, and with rocky and precipitous banks on either side. The length of the Strait is about $11\frac{1}{2}$ miles, its width of water-way varies from about 1,000 feet to three-quarters of a mile, and the average height of the shores on each side is above 100 feet.

For a long time the land traffic to Holyhead had been

---

requesting him to contribute the theoretical illustrations of the principles of the structure. The work was magnificently got up, partly at Mr. Stephenson's own expense, and has now taken an acknowledged position as one of the best standard engineering works of the present age.

In the present notice, the data given in that work are adhered to as closely as possible.

made to descend the bank, cross by a ferry, and ascend again on the other side ; but the inconvenience, loss of time, and often positive danger of this passage, prompted at a very early period efforts to establish a permanent roadway across the ravine. Bridges of timber or stone, embankments with drawbridges for the passage of vessels, and tunnels, had all been suggested; and as early as 1785 a petition was presented to Parliament for the means of carrying into effect one of these schemes; but the measure had not at that time assumed such an importance as to warrant the necessary large expenditure. When, however, Ireland was united to Great Britain, in 1801, the intercourse between the two kingdoms rapidly increased ; the inconvenience and danger to travellers were naturally and justly complained of ; and the attention of Government became seriously directed to the provision of a remedy.

They directed the late Mr. Rennie to survey the Strait, and he prepared four designs of bridges for crossing at different sites, the chief features of all being large iron arches, in some cases as much as 450 feet span, and 150 feet above the water.

Local opposition, however, and a disinclination to provide the large sum required, caused the postponement of the matter till 1810, when a parliamentary committee was appointed to enquire into the state of the roads from Shrewsbury and Chester to Holyhead. After taking much evidence, this committee reported that the whole subject, including the bridge, required further professional investigation ; and in consequence of their report, the Lords of the Treasury, in May 1810, instructed Mr. Telford to make an accurate survey of the roads, to

report on their improvement generally, and to consider
the best mode of passing the Straits.   He proposed two
plans of bridges—one, a single cast-iron arch of 500 feet
span, and 100 feet high; the other, a series of iron and
stone arches of smaller size—preferring, however, the
former.   These proposals were investigated in 1811 by
another parliamentary committee, who strongly recommended the execution of the large iron bridge.   The great
dimensions of opening could not be dispensed with.
The Straits, though tortuous and rocky, and of difficult
navigation, were yet constantly used by large ships, on
account of their sheltered situation, and the saving which
they afforded of about 60 miles extra journey round
the exposed and dangerous coast of the island; and
hence it was absolutely prohibited that any fixed structure should be thrown across, except of such width and
at such height as would allow the passage of large vessels
underneath without inconvenience or danger.

Notwithstanding the approval of the parliamentary
committee, still nothing was done, till a circumstance
that occurred elsewhere gave a new turn to the design.
In 1814 Mr. Telford was engaged in investigating the
possibility of throwing a bridge across the Mersey at
Runcorn, and finding the ordinary plan unavailable, he
had proposed a large bridge on the suspension principle,
which about that time was being brought into notice by
Captain Brown.   In 1815 a parliamentary commission
was appointed to carry into effect the various improvements required in the Holyhead roads, Mr. Telford
being appointed their engineer; and, after the general
road works had proceeded for about two years, the
enquiry arose whether the suspension principle might

not be advantageously applied to the crossing of the
Straits. Mr. Telford therefore again directed his attention
to the subject, and early in 1818 submitted a report,
design, and estimate, so strongly in favour of the suspension plan, that it was at once sanctioned by Government, and the works were put in hand in the latter part
of the same year.

This resulted in the well-known magnificent suspension
bridge, which, while it carries the road over the chasm
at a convenient level, offers an uninterrupted water passage
of nearly 550 feet wide and 120 feet high at high-
water, dimensions sufficient to allow the largest ships using
the Straits to pass under in full sail. Considering how
little experience had been gained at that time in the use
of iron for bridge construction, this bridge, so novel and
daring in design and so successful and elegant in execution, has conferred lasting and well-merited fame on
the engineer to whom its erection is due.

Telford's bridge was opened in 1826, but in a few
years after that time the new system of communication
began to supersede the ordinary roads. The metropolis
of England was soon brought into railway connection
with the great commercial port of Liverpool, and public
attention began to be directed to a similar improvement
of the communication with Ireland. A railway was projected for the land part of the journey; but, before its
direction could be decided on, a question arose as to the
merits of Holyhead as a point of departure, compared
with another port on the main land, somewhat further
south, named Port Dynllaen. Each of these had its
advocates as a packet station, and various investigations

were entered upon, and reports made, both by civil engineers and naval officers.

These discussions ended in a decision in favour of Holyhead, which led to the adoption of a line of railway to that port from Chester, to be connected by a branch with the Birmingham and Liverpool Railway at Crewe.   This line, called the Chester and Holyhead Railway, was first surveyed by Mr. George Stephenson about 1838, but was subsequently taken up and carried into execution by his son.

The Act was obtained (with a certain hiatus which will be hereafter referred to) in 1844, and the railway was opened for its entire length, including the passage across the Britannia Bridge, in 1850.

Few railways have exceeded this line, either in public importance or in engineering interest.   The natural difficulties have been great, and a series of engineering works of almost unrivalled magnitude characterise its whole length of $84\frac{1}{2}$ miles.   It emerges from Chester through a tunnel, and passes over a viaduct of 45 arches to the bridge by which it crosses the River Dee.   From thence it follows the embanked channel of this river and its estuary, and farther on the shore of the Irish sea, having here and there important works, until it is stopped by the bold headlands of the Great and Little Orme's Head. It then leaves the coast, and, passing through the narrow valley that separates these headlands from the main land, crosses the River Conway, beneath the castle walls, by a wrought-iron tubular bridge of 400 feet in one span. Passing through the town and under the walls by a short tunnel, it again reaches the coast at the Conway Marshes, and continues its course along the shore through the green-

stone and basaltic promontories of Penmaen Bach, and
Penmaen Mawr, the terminating spurs of the Snowdon
range, which it passes by two tunnels cut in the solid
rock.   Beyond these it is carried for some distance along
the beach, partly on a viaduct of cast-iron.   The sea
walls and defences on the one hand along this exposed
coast are all on a large scale ; whilst on the other side of
the line, a timber construction, similar to the avalanche
galleries on the Alpine roads, protects the line from
the *débris* rolling down from the lofty and almost overhanging precipices above.   The road again turns inland
to Bangor, and thence rises continually to a proper level
for crossing the Straits.   In this space it passes through a
very rough country.   The River Ogwen is crossed by a
viaduct 246 yards in length, and the Cogyn by one of 132
yards long and 57 feet high ; and three ridges of hills
are perforated by tunnels, 440, 920, and 726 yards in
length respectively, through hard primitive and trap
rocks.   In Anglesey the road passes over a marsh,
and through a tunnel 550 yards long, and enters Holyhead by partly making use of an embankment previously constructed by the commissioners for the turnpike
road.

When the Bill for the line was presented to Parliament
in 1843–4, the chief engineering work involved in it was
the bridge over the River Conway.   The passage of the
Menai Straits was proposed to be effected by permanently appropriating to the railway one of the two
roadways of Telford's great suspension bridge.   As the
strength of this bridge, however, was deemed inadequate for the safe transit of heavy locomotive engines, it
was intended to convey the trains across, in a divided

state, if necessary, by means of horse power, another locomotive being in readiness on the opposite side—the passage of engines being thus entirely obviated. The Commissioners of Woods and Forests, however, refused to allow a permanent appropriation of the half of the bridge in this way; and as the expense to be incurred was inconsistent with the idea of a *temporary* expedient, the Railway Company were driven to abandon this part of their plan, and to propose an independent bridge for their line. The Bill was accordingly passed with a hiatus of five miles at this part, to give time for the arrangement of the plans.

The directors at once instructed Mr. Robert Stephenson, who had then become their engineer, to select a suitable place for crossing ; and, after studying the subject well, he decided on a site about a mile to the west of Telford's bridge. The tide-way is here somewhat contracted ; but the feature which principally determined the choice was the existence of a rock or island in the middle of the stream, called THE BRITANNIA ROCK ; and from this, and not, as is often supposed, from any allegorical allusion, the bridge takes its name.

As the rock gave the opportunity of building a large pier, and so dividing the span into two parts, it was proposed to construct the bridge of two cast-iron arches, each 350 feet span, with a versed sine of 50 feet, the roadway being 105 feet above the level of high-water at spring-tides. The difficult problem of erecting these gigantic arches, in a situation where no centering or scaffolding would have been possible, was proposed to be solved by Mr. Stephenson in a very ingenious manner, and the Company prepared, at the end of 1844, a bill based upon this plan to go before Parliament the ensuing session.

As soon, however, as it became known what kind of a bridge it was proposed to build, a storm of opposition arose from the parties interested in the Straits, on the ground that such massive constructions would seriously interfere with the navigation. In March 1845, the Admiralty, in whom the guardianship of the navigation was vested, instructed three eminent engineers to examine the site and to report on the proposed plan; and as they stated that, in their opinion, the cast-iron bridge was ineligible, and that a clear passage of at least 100 feet high throughout the whole span should be insisted on, the proposal was abandoned.

Mr. Stephenson had already anticipated and prepared for this decision. He had fallen back upon the idea of the suspension bridge, and had begun to consider whether it was not possible to stiffen the platform so effectually as to make it available for the passage of railway trains at high velocities. His attention was directed to a suspension bridge at Montrose, where great stiffness had been afforded by a judicious system of trussing; and, carrying out this idea further, he conceived that sufficient strength might be obtained by the combination of the suspension chains with deep trellis trussing, having vertical sides, with cross bearing frames at top and bottom; the roadway being thus surrounded on all sides by strongly trussed framework. But as this idea was dwelt upon, difficulties arose about the material in which this trussed framework should be made. Timber was deemed inadmissible by reason of its perishable nature, and the danger from fire; and Mr. Stephenson, reverting to the

design he had made for a small bridge in wrought-iron
in 1841, was led to consider the application of this
material, by substituting for the vertical wooden trellis
trussing, and the top and bottom cross beams, wrought-
iron plates riveted together with angle iron. The form
which the idea then assumed was, consequently, that of
a *huge wrought-iron rectangular tube, so large that railway trains might pass through it*, with suspension chains
on each side.

The conception having reached this stage, only a little
farther careful consideration was necessary to arrive at
the idea that such a tube would, if properly designed,
serve the purposes of a *beam or girder*. The top and
bottom of the tube, which it was intended to compose
of thick wrought-iron plates, would evidently correspond
with the top and bottom flanges of a common cast-iron
girder, and might be made to perform their duties and
take their strains; and having reference to this consideration, Mr. Stephenson began now to regard the tubular
platform *as a beam, comprising in itself the main element
of its supporting power*, and to which the chains were
merely auxiliaries. Rough calculations were made, which
though necessarily very imperfect, gave confidence in
the feasibility of the design; and Mr. Stephenson's
reliance on it was further strengthened by some practical
examples brought to his notice of the great strength
shown by large iron vessels accidentally placed under
circumstances of peculiar strain and trial. Mr. Stephenson, fortified by these facts, even went so far as to propose
to dispense with the auxiliary chains altogether.

Thus the matter stood at the beginning of April 1845,
when the reports to the Admiralty put an end, as Mr.

Stephenson had anticipated, to the scheme of a cast-iron arch-bridge. The forethought and prudence with which he had prepared for this contingency, strikingly illustrate an element in his character, which was prominent through his whole professional career. Though he had strong confidence in his own views, when they were the result of sound reasoning and careful consideration, he never trusted with too sanguine an expectation to the favourable result of uncertain chances. He never undertook a doubtful course, without previously having determined a way of escape if it turned out contrary to his expectations; and to this admirable prudence is due, without doubt, much of the success which attended his professional labours.

The extinguishing of a favourite scheme, for such the arch-bridge was, would have damped the ardour of many men; but no sooner had it occurred than Mr. Stephenson announced to the directors of the railway that he was prepared to carry out a bridge of such a kind as would comply with the Admiralty conditions; and, after he had explained his views, they—not, however, without some misgivings—gave him their confidence and authorised him to lay his designs before Parliament.

The Bill came before the Committee of the House of Commons early in May. Mr. Stephenson's proposals, given on the first day, were received with much evident incredulity, and the Committee desired further evidence, and especially that of the Inspector-General of Railways, General Pasley, before they could pass the Bill authorising the erection of such a bridge as that which he had proposed. The Inspector concurred in the soundness of the idea, but most decidedly objected to the removal of the chains; and Mr. Stephenson, though he still expressed

confidence in the sufficiency of the tube alone, thought
it expedient to defer to this opinion, and to acquiesce in
their retention. This satisfied the Committee, and the
Bill in due course became law, by receiving the Royal
assent, the 30th of June 1845.

It was now necessary to take steps in earnest for
designing the tube. The calculations already made had
been very rough: for such constructions being entirely
novel, no experimental data were in existence of any use
for practical purposes. No wrought-iron beam of any
magnitude had ever been made or designed at all, and
though the general properties of the material had been
to some extent ascertained in suspension bridges, iron ships,
and other wrought-iron constructions, the way to apply
it in the best manner, so as to render its strength available
in forming a large girder, was quite unknown. It was
not the mere arrangement of the materials to resist the
transverse strains which formed the difficulty of the
problem. It was rather the practical design of any such
structure at all—the difficulty of obtaining the iron in
the forms required, or of adapting such forms as were
obtainable to new purposes—and of devising a beam, not
merely strong enough for its ultimate use as a bridge, but
of sufficient independent rigidity for keeping its form
when erected, and for sustaining the complicated and
trying processes connected with its first construction, its
floatation, its conveyance to the site, and its elevation and
fixing in place.

Mr. Stephenson, therefore, considering the magnitude of
the matter at stake, at once decided on supplying the
want of data by a series of experiments on a large scale,

before committing himself to the design for the tube. His own knowledge of the properties and manufacture of iron was very considerable, having been engaged from his youth up so actively and prominently in the manufactory at Newcastle; but, with the unassuming modesty of true merit, he did not think fit to rely on himself alone, for he felt that, considering the responsibility which he had publicly assumed, he would be doing injustice to the Board of Directors, who had placed such confidence in him, if he did not avail himself of all the practical and scientific aid within his reach. He accordingly entrusted the performance of the experiments to Mr. William Fairbairn, of Manchester, whose practical experience he estimated very highly, and with whom he had consulted on the subject previously to the parliamentary investigation. A short time afterwards, also, at Mr. Fairbairn's suggestion, he engaged the assistance of the late Mr. Eaton Hodgkinson, whose valuable contributions to engineering science, more especially in regard to iron structures, had attracted much notice in the profession.

The experiments, which were designed and proceeded with under Mr. Stephenson's personal superintendence, were not at first specific in their object. It was necessary rather to determine what kind of information was required, than to pursue any definite course—to ascertain generally in what manner tubes might be expected to fail, and to what extent their strength might be modified by form. The first idea of the tube was a rectangular section, but this was afterwards thought objectionable, and attention was directed to the circular or elliptical form. Model tubes of these sections were accordingly made and carefully tested; but they failed in strength, and, after due discussion and consideration of the experiments, it was decided that these shapes were ineligible, and the original rectangular form was reverted to. Trial tubes of this shape proved more satisfactory, and, in February 1846, Mr. Stephenson was able to report to the half-yearly meeting the general conclusion that had been arrived at. He stated that, after carefully studying the results as they developed themselves, he had satisfied himself that the wrought-iron tube was the most efficient as well as the most economical description of structure that could be devised for crossing the Straits—that the form of the tube must be rectangular—that the general disposition of the material had been determined—and that the only problem remaining was that of the necessary strength to be given : that apparently greater strength was required than had been at first proposed; but to establish the formulæ of calculation more positively, as well as to settle doubtful points regarding the use of the chains, it was desirable to carry the experimental researches still further.

It was in this preliminary series of investigations that the remarkable and unexpected fact was brought to light that the power of wrought-iron to resist compression was much less than to resist tension, being the reverse of that which held in cast-iron. This discovery had not only an important bearing on the design for the tube, but it has since formed a valuable datum in regard to the engineering use of the material generally.

Another point having important influence on the subsequent design was also brought out for the first time. It was found that in all the earliest trials of thin tubes the top part, which was exposed to a compressive strain, failed not by the direct crushing of the material, but by the buckling or collapse of the plates. This was a new fact altogether, and one which had never been taken into account in any of the formulæ for strength previously employed. It indeed annihilated at once their practical utility; and, prominent as it became in subsequent experiments, it threatened temporarily even to frustrate the consummation of Mr. Stephenson's design. It was, therefore, at once treated as the most important object of investigation. In some of the elliptical tubes a sort of cell or fin was introduced; but as this form was just then abandoned, the same difficulty arose with the rectangular tubes, the tops of which, when formed of thin flat plates, buckled up under the pressure. At length corrugations were made in the plates, which were found to add much to the stiffness; and this led to the formation of the top in a series of tubes or cells, which, while they gave the necessary rigidity, offered great facilities for the manufacture, as well as convenient access to all parts of the material; an object which had been always prominent in Mr Stephenson's mind.

The publication of Mr. Stephenson's report on these preliminary experiments, which was accompanied by others from Mr. Fairbairn and Mr. Hodgkinson, formed an important epoch in the history of the bridge. Public attention was now for the first time drawn to the subject, and the directors of the Company were relieved from some anxiety by the more definite details submitted to them. But still the necessity for further experiments was obvious. Everybody had some doubts and fears to suggest—dismal warnings came in on all hands, suggesting every imaginable apprehension. The necessity for chains was still advocated, not only by General Pasley, but by Mr. Hodgkinson himself. Many doubted the efficiency of riveting to unite such a mass of plates; some foretold the most fatal oscillation and vibration from passing trains, sufficient even to destroy the sides of the structure; while others asserted the insufficiency of the lateral strength to resist the wind. In fact, with few exceptions, scientific men generally either remained neutral or ominously shook their heads and hoped for the best, and even the most sanguine waited for further experimental investigation. All this was so discouraging, that Mr. Stephenson, strong as his faith was in his own plans, could not avoid appearing at times disheartened, when he withdrew from the turmoil of his metropolitan parliamentary duties to deliberate on the weighty difficulties he had to encounter in his gigantic undertaking in the distant hills of North Wales.

At this time, too, another serious matter presented itself. The preliminary considerations, discussions, and experiments summed up in Mr. Stephenson's report had occupied much longer time than had been anticipated; but the work on the other portions of the line had been steadily progressing, and it became evident that the Britannia and Conway bridges would be ultimately the chief cause of delay in the completion of the line. Hence the directors became impatient that Mr. Stephenson should sufficiently mature his plans to allow of the commencement of the masonry; and, while they did not hesitate to sanction the continuance of such further experiments as he might deem advisable, they, with a confidence in his proposals which few shared with

them, entreated him without delay to commence operations simultaneously at both sites, and to complete
his designs as he proceeded. This gratifying resolution
added considerably to his anxiety, as he wished first
to complete the smaller structure—the Conway Bridge,
in order to obtain for the larger one the benefit of any
experience it would afford. The plans of the masonry
were however at once prepared. They were ready for
contract by the middle of March 1846, and the first stone
of the Britannia Bridge was laid April 10 in that year.

The further experiments which were needed for the
completion of the design of the tubes, were of two
kinds. In the first place it was considered necessary
to make a model tube, very much larger than any
of the previous ones, and representing more nearly
the principles of the structure itself; with a view of
putting it to every possible test, and by constant correc-
tion of its weak points of arriving gradually at the best
form and proportions possible. And, secondly, it was
found that, in order to give the power of reasoning from
this model up to the structure itself, many more experimental data were necessary than were yet in existence, as
to the qualities of the materials and the workmanship
proposed to be used, and the influence of strains upon
them.

These latter specific enquiries were undertaken by Mr.
Hodgkinson. They consisted of careful and elaborate
experiments and deductions on the compression, flexure,
and crushing of materials and manufactured compound
structures under direct pressure—on the extension and
tensile strength of materials—on riveting—on the shear-
ing of iron exposed to transverse strain—and on the

transverse strength of beams and tubes of various kinds.
They were, it is true, more particularly aimed at the question then pending; but they form a mass of general information of the most useful description, and probably, as a whole, unrivalled in extent and value.

The large model was constructed at Mr. Fairbairn's works at Millwall, near London, in order that it might be tested under Mr. Stephenson's more immediate inspection. The proportions having been thoroughly discussed, it was commenced in April 1846, and completed in July, and the experiments upon it were immediately put in hand. It was rectangular in shape, with a top composed of one row of cells, and its dimensions were determined in reference to the requirements for the Britannia Bridge, every dimension being one-sixth of the eventual magnitude then thought necessary. Thus the Britannia tube being 450 feet long in the clear, the length of the model between the bearings was 75 feet—the depth 4 feet 6 inches—and the width 2 feet 8 inches: forming a large bridge-girder of itself. The weight was between five and six tons. It was supported at each end on a pier, and weights were hung on the centre till it gave way. In the first experiment it broke with $30\frac{1}{4}$ tons by the rending asunder of the bottom plates. These were then repaired and strengthened, when it bore 43 tons, giving way at the sides, which were then strengthened in turn. Next the bottom gave way again several times, each time having larger dimensions; and so the trials and alterations were continued until at length a proportion was arrived at which proved to be about equally strong all over. As thus perfected, the tube bore 86 tons, or $2\frac{1}{2}$ times that of the first trial, although in the strengthening only one ton of iron had been added—such being the effect of a judicious application of the material.

The experiments on this model directly proved what at first had appeared problematical, namely, that with such extensive horizontal developement of the top and bottom flanges, the whole of their sectional area would act effectually in resisting extension or compression throughout the entire width. In fact, when the model beam was broken, the tearing asunder of the bottom plates actually commenced at about the middle of the tube, and not at the outside edges—showing thus that the principles of simple girders were directly applicable to this construction also.

The experiments on the large model were continued till April 1847 ; but in the meantime the designs for the tube had not been neglected. During the first half of the year 1846 a great number of tentative drawings, models, and calculations were made; and although many of these attempts were necessarily discarded, as clearer views resulted from increased experimental information, yet some of the designs thus sketched out remarkably anticipated the ultimate plan. In July, when the experiments on the large model were commenced, a design for the great tubes had been made out in considerable detail. This was gradually improved as further information was obtained ; and more perfect drawings were completed in the beginning of November. These, however, were further modified from time to time, the most important change being in the arrangement of the cells of the top, effected in February 1847, in accordance with certain principles resulting from the enquiries of Mr. Hodgkinson.

In March the correct lists of the plates were made out, and the first complete working drawings for the tubes were finished, although still further improvements were introduced as the work went on.

There is little doubt that this gradual growth and constant improvement of the designs conduced much to their perfection; and at a much later period, when wrought-iron girders had been greatly developed by the experience of subsequent years, and the talent of engineers had given rise to numberless elegant and ingenious practical combinations in bridge construction, Mr. Stephenson declared that he found it difficult to conceive any better means than those adopted of solving the problem.

While the design of the tubes was thus being considered, another question of scarcely less importance had also called for investigation, namely, the means by which they were to be placed in their position. For it scarcely need be remarked that the immense size and weight of the tubes, and the peculiarities of the situation, put them completely out of the range of all ordinary experience.

Many suggestions for this purpose were made and discussed at various times. An early idea of Mr. Stephenson's, when the cast-iron arch-bridge was proposed for Conway, was to float it to its place on pontoons; and the merits and difficulties of this plan had been fully discussed. When the arch was superseded by the tube in both localities, this mode of placing the tube was again considered, as the form which the bridge had assumed was evidently favourable for such an operation; and it was accordingly proposed to construct the tubes on the beach, and to float them to their ultimate position. This presupposed sufficient strength in the bridge independently of chains; but Mr. Stephenson, at that time, considered that the insurance afforded by chains against any accident from unforeseen causes would be a consideration of vital importance; and he did not, in that stage of his experience, feel justified in throwing away such a security. He therefore determined on availing himself of these auxiliary suspension chains, in the first instance, for supporting a temporary platform or scaffolding, along which the tubes constructed on the land could be rolled into their places. This plan was maturely considered; the designs for the platform,—which would of itself have been a large suspension bridge—were prepared; and much attention was bestowed on the manner of making the chains available as additional means of security to the tube, after their temporary office as scaffolding had been performed.

As, however, the progress of the design in the early part of 1846 gave more confidence in the self-supporting power of the tubes, and as the completed estimates for the suspension platform, with the then high price of wrought-iron, were very large, the subject was again discussed; and in July, Mr. Clark, who had accidentally obtained what he considered a good practical suggestion of a mode of raising the tube, urged upon Mr. Stephenson, with Mr. Fairbairn's assistance, a renewal of the floatation scheme. The subject was carefully and candidly reconsidered by Mr. Stephenson, and ultimately the chains were abandoned; and it was decided to put the tubes together upon the shore of the Straits; to float them to their site on pontoons; and to raise them to the required

high level by hydraulic power; and this was the plan
carried into practice.

Meantime it became urgent that arrangements should
be made for the manufacture of the tubes, which the
directors decided to put out to contract, reserving to
themselves, however, the right to purchase the iron,
and to supply it to the makers of the tubes at a fixed
price per ton. In July 1846, the plates were contracted for by seven of the principal iron makers in the
midland iron district; and shortly afterwards negotiations were commenced with several manufacturers for
the construction of the tubes, but it was May 1847,
before the arrangements were finally concluded. The
first stipulation had been that the makers should construct the work at their own manufactories, in large
sections, to be delivered on the shore of the Straits, and
there put together; but as this plan involved difficulty,
it was afterwards decided that the manufacture should
be entirely done on the shore. On this understanding
the contract for one large tube was given to Messrs.
Garforth, of Manchester, and for the other seven to Mr.
Charles Mare, of Blackwall. The site for the construction of the tubes had been determined some time
previously. It was necessary that the making of the
four large tubes should proceed simultaneously, and the
clearing and preparation of the four places where they
were to be made was a work of considerable difficulty
and labour. Large and strong platforms of timber had
to be laid down to build the tubes upon: these were
occupied, and the ironwork began to get into shape by
July 1847, and the first rivet for putting the tubes
together was inserted on the 10th of August following.

The first of the large tubes was finished, and the wood platform was removed from beneath it by the 4th May 1849, leaving the weight of the tube supported on its two ends. This had, however, been anticipated by the Conway tube, finished in the January previous; and as the latter was in reality the first practical test of the great experiment, the anxiety of all concerned was intense to see the result as the timbers were gradually cut away. A deflection of 2 or 3 feet had been predicted, and many high authorities had affirmed that the tube could not support its own weight; while others foretold the buckling of the top, the distortion of the sides, the crushing of the extremities, and all sorts of failures. These forebodings were set at rest, and all fears at an end, when the platform was cleared away, and the tube took its own weight with just about the calculated deflection, and without the slightest appearance of undue strain or damage in any part.

The second tube was finished a few weeks after the first, the third in October 1849, and the fourth in February 1850.

Two other operations yet remained, each as gigantic and novel as the construction of the tubes, and attended with as much anxiety; namely, their removal by floating from the shore where they were constructed to the site of the bridge, and the hoisting of them up to their required level.

In the Conway Bridge these operations had been undertaken by the contractors; but, in the more important case of the Britannia Bridge, Mr. Stephenson preferred that they should be done immediately under his own direction.

The arrangements for them had accordingly occupied his
attention during the latter part of the year 1848, and to
facilitate the study a model was made of considerable
size, with real water, on which the whole operation could
be imitated; so that by constant rehearsals of the process
on this miniature pool the plans for floating were matured.
Each tube was to be floated on eight pontoons, introduced in cuttings in the rock under the tube, and which,
on a certain day, were to be emptied and allowed to rise
by the flowing tide till they lifted the tube off its bearings,
and took its weight upon themselves. They were then
to be hauled out into the stream, in order that it might
float them and their burden to the bridge, being carefully
guided and controlled by hawsers attached to the shore
on either side. This was to be done near high-flood,
so that the tube might arrive at the bridge about the
turn of the tide at still water, when its ends were to
be lodged upon shelves prepared for the purpose at the
foot of each tower, and the pontoons floated away. The
difficulties of this operation consisted in the magnitude of
the moving mass—the great number of departments and
of hands entering into the process—the short time it had
to be done in (only about one hour and a half)—the
great velocity of the tide (about six miles an hour)—
and the terrible consequences that might ensue if the
operation should fail, and the floating mass become
unmanageable under the swift and powerful ebb-tide.
Mr. Stephenson's energy, prudence, and foresight were
here again admirably displayed. He devoted untiring
attention to the organising and teaching of a large body
of persons, many hundreds in number, who were to
be engaged in the task, directed by his own staff

of assistants; and as the work involved operations of
a nature new to engineers, he obtained the aid of a
large body of sailors and nautical men, under the command of Captain Claxton, R.N., who had acquired much
reputation for his successful exertions in rescuing the
Great Britain, stranded at Dundrum Bay. And further
than this, impressed as Mr. Stephenson was by the immense responsibility of the operation, he invited two
of his most eminent brethren in the profession—now,
alas! like him, departed from this earthly scene of
their labours—Mr. Brunel and Mr. Locke, to give him
the benefit of their assistance, a trait of professional good
feeling which did him infinite honour. This aid, we need
hardly say, was cheerfully afforded, both gentlemen being
at his side the whole time.

The floating of the two Conway tubes in March and
October 1848, had served as useful preliminary trials,
from which much valuable experience had been gained,
and which enabled Mr. Stephenson well to mature his
plans. Preparations were made for floating the first
Britannia tube on the 19th of June 1849, but in consequence of the fracture of a capstan at the commencement,
it was postponed to the next day, when it was successfully performed—the lodging of the tube upon the
shelves of the towers being greeted by cannon from the
shore, and the hearty cheers of many thousands of spectators, whose sympathy and anxiety had been indicated
by the unbroken silence with which the whole operation
had been observed.

The tube lay across the water, out of reach of the
tide, during the remainder of June and the whole of July,

while the machinery for raising it to its proper height was fitted in the towers. This apparatus consisted of huge hydraulic presses, placed at the tops of the towers on each side, from which strong chains hung down to the tube. By working these presses, the tube was raised six feet at a time, the ends sliding up in recesses or vertical grooves, which were built up under the ends of the tube as fast as it rose, timber packing being further inserted, so that in case of fracture of any of the suspending machinery the tube would not have far to fall.

On the 10th of August the raising was commenced, and it proceeded slowly till the 17th, when one of the press cylinders burst, allowing the end of the tube to fall 8 or 9 inches on to the packing below—which, slight as the fall was, caused some damage. By the 1st of October the press was again ready : the raising steadily proceeded, and on the 13th the tube safely attained its final elevation.

The second tube was floated on December 6th, and it was in its place by January 7th, 1850. This was in a line with the first one, and, as the two short or land tubes corresponding were already completed, it only required the four lengths to be joined in order to effect a passage across the Straits. These junctions proceeded night and day, and were completed and the rails laid by March 4th. The next day Mr. Stephenson and some friends passed through on a locomotive, followed by an enormous train of forty-five coal wagons and carriages, containing seven hundred passengers, and drawn by three engines. On the same day the last rivet was formally put into the tube by Mr. Stephenson and the contractor, and the passage of the Menai Straits by the Chester and Holyhead Railway became an accomplished fact. On the 15th the bridge was passed by the Government Inspector, and on the 18th it was opened for public traffic. It was worked as a single line for some time. The third tube was floated on June 10th, an operation which the concurrence of several accidents made the most hazardous of all; and it was raised July 11th. The fourth and last tube was floated July 25th, and placed in position on August 12th. The last piece of scaffolding was removed on October 11th; and on October 19th, 1850, the bridge was completed and opened for the double line.

The description of the bridge need only be very brief, as full particulars and views have been made so accessible by publication. It will be confined to an enumeration of such prominent points in the structure as may best illustrate its novelty and magnitude.

The nature of the ravine over which the bridge forms a passage has been described, and the peculiar conditions of navigation of the Straits, have already been alluded to as having influenced the general design. The water-way was about 1,000 feet wide, with a rock in the middle, so that by building a tower of sufficient height on each side, and one on the rock, this space was divided into two equal spans. But as the shelving shore on each bank gave a considerable increase of width at the required level of the roadway, some mode was necessary for filling in this additional space. The simplest plan would have been to build it up from the ground with arches of masonry, as Telford had done in the Menai Suspension Bridge; but Mr. Stephenson resolved to make use of these side spaces to effect an important object in regard to the large tubes, namely, to diminish the strains upon them by making them parts of a *continuous* long beam, instead of leaving each a single isolated span. It is well-known that when a beam extends continuously over several openings —as, for example, in a floor-joist—the strain is much less than when each span is covered by an independent beam of the same size. This, therefore, was the principle which Mr. Stephenson put in practice in this case. He threw the abutments, or land terminations of the bridge, high up the rocks on each side, and filled in the land spans with shorter tubes, so that the bridge became one of four spans —two large ones in the middle, flanked by a small one on each side. As regarded the bridge itself, these smaller land tubes were not required at all: they merely acted, so to speak, as counterpoises for the large tubes, converting them into continuous long beams, and their overhanging weight serving to relieve the centre parts of a portion of their strain. This application of the principle of continuity is a good example of Mr. Stephenson's excellent intuitive practical perception of mechanics. The general fact was, indeed, known, and its explanation had been investigated in mathematical works; but it was not till long after the erection of the Britannia Bridge that it was brought prominently before the notice of the engineering profession, or applied to iron bridges generally, with any view to the advantages afforded by it.*

---

* See Minutes of Proceedings of the Institution of Civil Engineers, vol. ix. 1849–50.

It does not appear to have formed any important part in the preliminary experiments, or even to have been the subject of any recorded calculations. In all probability its application was dictated almost entirely by Mr. Stephenson's practical judgment, and the test of elaborate mathematical analysis subsequently applied to the work shows how sound and accurate this judgment was.*

The span of each of the long tubes is 460 feet clear of the towers—that of each of the short or land tubes, 230 feet. A separate line of tubes is provided for each line of railway, with a small space between them, but both resting on the same towers. The four land tubes were constructed *in situ*, upon scaffolding built temporarily for the purpose.

Each line of tubes is connected throughout, forming one continuous tube 1,511 feet long, and weighing, with the permanent way, 5,270 tons. This long tube is securely fixed in the centre tower, but its bearings on the side towers and abutments are moveable, that it may expand and contract freely from changes of temperature.

The depth of the tube externally is 30 feet at the centre tower, diminishing to 23 feet at each end, so that while the bottom outline is straight, the top forms a portion of a curve. The internal clear height at the ends is 16 feet 4 inches. The breadth of the tube is about 14 feet, allowing room for a man to stand safely on the side during the passage of a train.

As each span of the tube had to bear its own weight

---

* See Mr. Clark's work, p. 785.

between the supports before they were connected together, it was necessary, in the design, to treat each as a separate beam. The top and bottom members were the effective portions in resisting the strain, and in them, consequently, the largest amount of material was collected, being disposed in the shape of a series of square cells or flues, eight in the top and six in the bottom, of sufficient size to allow workmen to enter for the purpose of riveting, and also to cleanse and paint the interior. The sectional area of solid metal in the top, at the middle of the length of the large tube, is 648 square inches, of the bottom 585 square inches. This is reduced towards the ends.

The engraving, fig. 6, represents a section of the tube, and will give a general idea of its construction.

The sides are plain sheets of plates, stiffened by vertical ribs or pillars of T iron, within and without, and also by gussets or corner-pieces, filling up the angles on the inside. The sides increase very much in thickness towards the towers, and are strengthened at the end with massive cast-iron frames.

The entire weight of ironwork in the bridge is 11,468 tons—the rivets in the tubes number above two millions.

The strength of the tubes has been well determined by several modes of calculation. Considering one of the large tubes as an independent beam, it is found that it would not break with less than about 5,000 tons equally distributed along its length. Now the tube itself weighs 1,550 tons, and adding to this the greatest moveable load that could possibly come upon it would make up little more than 2,000 tons, or two-fifths

of its ultimate strength. But this is less favourable
than the reality, inasmuch as the strength is nearly

FIG. 6.

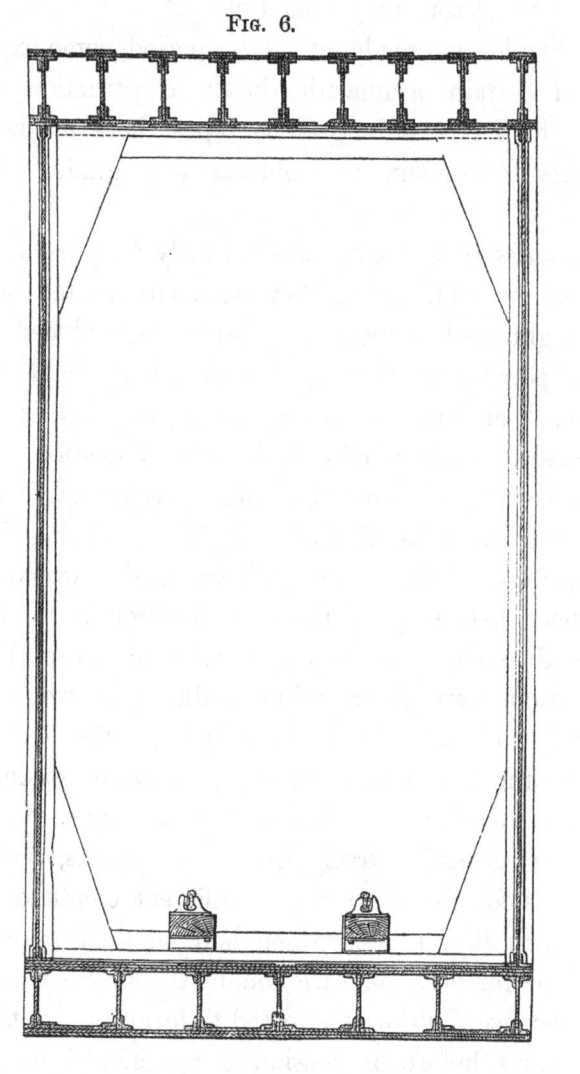

doubled by the *continuity* of the þeam over the several
spans.

The strain upon the metal at the middle of the length

of the long tube would be about $5\frac{1}{4}$ tons per square inch, if considered as an independent beam, but is reduced to $2\frac{3}{4}$ tons by the continuity.*

Mr. Stephenson made, at a later period, some explanations of certain peculiarities in the construction of the tubes which it may be well to repeat here, as they are necessary to explain the objects that guided him in the design.

The *sides* of the tubes weigh nearly forty per cent. of the whole weight. Had they been constructed *in situ*, this weight might have been considerably reduced. But in the operation of floating, the tubes were liable to be supported at any point of their length, besides being subjected to chances of considerable dislocation, and to disasters which, on more than one occasion, did actually threaten their entire destruction: The stiffening frames and gussets, which in an ordinary girder would have only been necessary at the ends, became therefore requisite throughout the whole length; and even the top and bottom were considerably modified, as while overhanging the pontoons at each end to the extent of 70 feet, the top, instead of being in compression, was thrown into extension. Again, the tubes had to be raised by being suspended freely from four chains, requiring provisions for this support of a different character from that which they needed when laid on their permanent bed; and further, the variation in the strains when the four tubes were ultimately joined to form one continuous tube—parts before in tension being then thrown into

---

* The calculations for determining this, furnished by the author of this chapter, will be found in Section viii. chap. 3, of Mr. Clark's work.

compression, and *vice versâ*—required a suitable arrangement of the material : the effect of all these provisions
being to increase the quantity and modify the arrangement of the metal in the tube. In proportioning, therefore, the parts of a· structure destined for such usage, the
mere consideration of the strain to which, as an ordinary
beam, it would be subjected, formed but a part of the
problem, and no fair direct comparison can be made
between the weight of this bridge and that of an ordinary
beam.

Mr. Stephenson was of opinion that some misapprehension existed on the object and importance of the cells of
which the top and bottom of the tube were composed, as
well as on the choice of form of the tube ; and he has given
explanations to clear up both these points. He shows
that to collect the necessary quantity of material of the top
and bottom in single plates would have required the
former to be 2·7 inches, and the latter 2·3 inches thick ;
and had such plates been procurable, nothing better could
have been desired, and the cells would have been unnecessary.

At that time, however, it was impossible to procure
plates of such a thickness, whose quality could be depended on ; and the engineer in this, as in numberless
other details, had to adopt what he could obtain. Now,
the arrangement of the plates in cells is almost the only
conceivable arrangement possible for getting the required section, allowing access, at the same time, to
every part for construction and future maintenance.
This alone led to their use in the bottom of the tube,
where the form was quite indifferent. With respect to
the top, however, it was of great importance, since

thick plates could not be had, to ascertain the best form
of cell for resistance to compression that could be devised with thin plates.   A series of valuable experiments by Mr. Eaton Hodgkinson led to the rectangular
cells actually used, not because such form presented
any peculiar advantage over any other, as some have
imagined, but because these experiments demonstrated
that cells of that magnitude and thickness were independent of form, and were crushed only by the actual
crushing of the iron itself.   Under these circumstances
the square cells were used as the best practical method
of obtaining the sectional area required.

Similar misapprehension has also existed as to the
considerations which led to the rectangular form of the
tubes themselves.

The result of direct experiments made with round,
oval, and rectangular tubes—there being precisely the
same section and weight of metal in all three—was
that the circular tube was the weakest and the oval
tube the strongest, the rectangular form being intermediate.   The oval tube was first studied with a view to
adoption.   Its form, however, was not favourable either
for its practical construction, or for its connection with
the suspension chains, which were originally intended to
be used in the erection; and practical considerations in
this case also dictated the use of the rectangular tube.   It
must also be remarked that the result of experiments
made on oval, round, and rectangular wrought iron tubes,
when reduced to the same depth and compared, was in
favour of the rectangular form—although, within ordinary
limits, the form was not proved to be a matter of great
importance.

The centre or Britannia Rock tower is 230 feet high. The base is 60 feet by 50 feet, and the size at the level of the tube is 55 feet by 45. The pressure on the base is 16 tons per superficial foot.

The side towers are 18 feet lower than the Britannia tower; the base of each is 60 feet by 37 feet; the size at the level of the tube 59½ by 36½ feet. The great height of the towers above the tubes was necessary for fixing the hydraulic presses which raised these ponderous masses into their places.

The shore abutments are 35 feet lower than the side towers.

The internal work of the towers and abutments is of Runcorn sandstone, with some brickwork. The exterior is faced with Anglesea marble, from quarries in the carboniferous limestone at Penmaen, the northern extremity of the island.

The total quantity of masonry in the bridge is nearly a million and a half cubic feet.

The design of the bridge, as regards its architectural character, must, considering the entire novelty of the form, and the colossal dimensions of the structure, have been an arduous thing to attempt. The object aimed at was the adoption of such a character as would best accord with the tubes, the external appearance of which is simply a representation of beams of gigantic proportions. With this view, a combination of the Egyptian and Grecian styles was thought the most appropriate, the former as applied to the general and more massive portions of the design, and the latter to the less ponderous parts and to the details generally. The colossal lions on the approaches, designed and executed by the late Mr. John

Thomas, were intended as allegorical representations of the strength of the edifice and the boldness of the undertaking.

A colossal figure of Britannia was designed also by Mr. Thomas for the centre tower, but the great expense prevented its construction.

It is unfortunate that the bridge consists of an even number of spans, architectural beauty requiring an opening in the centre and not a pier. But the existence of the rock which determined the site of the bridge left no option on this point.

The total cost of the Britannia Bridge was a little over £600,000. The ironwork cost £375,000 or nearly £33 per ton—a very high price, no doubt; but it must be recollected that at the time these contracts were made iron was very dear, and the character of the work was new. At the present day there would be no difficulty in getting it for about half the sum.

The cost of the experiments was about £5,300.

Since the bridge has been in use the deflection has been carefully tested from time to time, and no perceptible increase has taken place. The painting has been attended to, and the tubes have been covered by a roof to shelter the ironwork from the rain. Mr. Stephenson continued to satisfy himself as to the condition of the bridge until near his death, and gave the opinion that he found it difficult to conceive that even the lapse of centuries could in any way affect such a structure.

It is to be hoped this opinion may be borne out by experience, and that the bridge may prove one of the most durable, as it certainly is one of the most remarkable, monuments of the science and enterprise of the present age.

Conway Bridge.

A few words must be added relative to the Conway Bridge, which has been mentioned incidentally in the account of the larger structure. The difficulties here, also, were formidable. It was necessary for the railway to cross the Conway River, a large estuary running high up into the land. The average width is about three-quarters of a mile, but advantage had been taken by Mr. Telford of a rocky island intercepting the channel, to reduce the width to a much smaller space, which he spanned with a suspension bridge for the passage of the Holyhead road. There could be no doubt that the proper site for crossing with the railway must be close to that occupied by the road; but it was also evident that, on account of the great depth at this point, 63 feet at high water, and the fearful velocity with which the tide ran through it to fill the large expanse above, it would be impossible either to build any intermediate pier in the water way, or to fix any centring or scaffolding for the erection of the bridge. The span of the suspension bridge is 315 feet, but from the form of the rocks the least span that could be obtained alongside it was 400 feet, and thus the problem became, to erect a bridge of this width in one span and without any fixed scaffolding. It will be recollected that Mr. Stephenson's first idea for the Menai Straits was to construct the bridge of large cast-iron arches, and it was proposed to treat the Conway river in a similar way, one colossal arch spanning the entire opening. The principal difficulty was with the erection; the ingenious plan which Mr. Stephenson had contrived for the Britannia Bridge was inapplicable here, and he proposed to build the arch upon pontoons, which, when the work was finished, were to be floated to the site, and made to deposit the entire structure at once upon its bearings.

When, however, the arch scheme was abandoned for the Menai Straits, it was also put aside for the Conway, as it soon became apparent that the problem was essentially so identical in the two cases, that any design adopted for the larger structure would, in all probability, be the most suitable for the smaller. Hence no further special attention was given to the Conway crossing till the general principles of the Britannia Bridge were settled, after which the two designs progressed simultaneously.

It was, however, at the Conway Bridge, as has been already stated, that the success of the great experiment was first put to the test. The contract for the tubes was let in October 1846 to Mr. Evans, who was already executing the masonry. It was this enterprising man who first proposed to construct the tubes entirely on the site, a plan afterwards adopted at the Britannia Bridge with so much advantage; and in the case of the Conway he undertook, at his own risk, the arduous and perilous duty of floating the tubes and of erecting them complete in their places, which he accomplished very successfully. The contract price paid to him for the tubes fixed complete was only about £4 per ton more than was given for the tubes only at the larger structure. The first stone of the bridge was laid May 12, 1846, but the manufacture of the tubes was not commenced till March 1847; the first tube was tested in January 1848, floated to its place in March, and ultimately raised and in use for railway traffic in April, a rapidity of execution highly praiseworthy. The second tube was floated in October, and the bridge was opened for traffic on the double line in December 1848.

There are two tubes, one for each line of railway; they are 400 feet long in the clear between the supports; the external height in the middle of the length is 25 feet 6 inches, diminishing to 22 feet 6 inches at the ends. The height of the bottom of the tube above high-water line is 17 feet. The general design of the tube corresponds with that in the Britannia Bridge, but the arrangement of the material is somewhat modified, from the circumstance that the latter is designed to act as a continuous beam, whereas the former is an independent one. The tubes were constructed on the shore of the estuary above the bridge, floated down to the site on pontoons, and raised by hydraulic power, as in the Britannia Bridge.

The artistic design of the Conway tube will probably be considered less successful than that of the Britannia Bridge; the situation is picturesque in the highest degree, and the elegant suspension bridge rather added to than diminished its beauty; but the same remark will hardly apply to the subsequent erection.

An attempt was made to give a style corresponding to that of the castle, but alterations subsequently introduced into the construction, and the omission of the ornamental parts to save expense, crippled the design; and the circumstance of the tubular bridge not being parallel to, but considerably askew from the suspension bridge immediately alongside, is a sad eyesore.

The unfettered reign of private enterprise, which, under the dictatorship of the engineer, has of late so much prevailed in this country, has been no doubt a grand source of works of commercial utility, but it has doomed us to much bitter humiliation in matters of art and taste.

W. P.

CHAPTER IV.*

THE HIGH LEVEL BRIDGE AT NEWCASTLE-ON-TYNE.

Object of the Bridge—Ravine of the Tyne—Ancient Bridge at a Low Level—Inconveniences of the Passage—Early Proposals for a High Level Bridge—Mr. Green's Scheme—High Level Bridge Company—Mr. Stephenson appointed Engineer—Newcastle and Darlington Railway—Proposal for the Double Roadway—Parliamentary Proceedings—Description of the Bridge—The Piers—The Iron Superstructure—Mr. Stephenson's Motives for the Adoption of the Bowstring Girder—Letting of the Contract—Driving the Piles—Manufacture of the Ironwork—Erection—Completion.

THIS bridge, although of less magnitude than either of the two other large iron bridges selected for illustration, is one of the most celebrated of Mr. Stephenson's works. It has for its object to form a double communication, by railway and by common road, at a high level, between Newcastle on the north, and Gateshead on the south bank of the river Tyne.

The river runs through a deep valley or ravine, the average level of the land on each side being about 100 feet above the water of the river. A bridge of considerable antiquity, crossing at the bottom of the valley, formed the only passage, and as the streets leading down from the level part of the town on each side were exceedingly steep, the passage from one elevated shore to the other was

---

* This chapter is contributed by Professor Pole.

fraught not only with much difficulty but with positive danger.

This, however, was endured for hundreds of years as a necessary evil; it was only about the commencement of the present century that the idea of avoiding the difficulty by a bridge at a higher level began to be seriously entertained. The matter was first mooted by Mr. R. B. Dodd, a local engineer, and a public meeting was held in furtherance of the plan, which, however, did not meet with any adequate support.

About 1825 Mr. Telford proposed a bridge on the present site, and other engineers are said to have renewed the scheme from time to time; but the project best known was proposed, some years after Mr. Dodd's, by the late Mr. John Green, an architect of Newcastle, and an attempt was made to form a company to carry it out, the chairman being Mr. John Hodgson Hinde, an influential inhabitant of the town. The late Lord Grey and several other gentlemen of local influence exerted themselves to promote the measure, but the requisite capital, £30,000, was not forthcoming from the public; while the corporation of the town felt they were not warranted in incurring an enormous outlay, which could not, they conceived, be reimbursed by any tolls they could impose upon the traffic across it.*

Mr. Hinde, however, still persevered. In 1843 he induced Mr. George Hudson to aid in the undertaking, and to become vice-chairman of the company, and it was resolved that Mr. George Stephenson should be consulted on the

---

* The tolls on the carriage road only of the present bridge amount to above £4600 annually, or 15 per cent. on Green's estimate.

matter. Accordingly a new prospectus was drawn up of the 'High Level Bridge Company,' in which his name appeared on the direction, while that of his son was appended as consulting engineer. George Stephenson had a design prepared of a bridge with bowstring girders and a double road, similar in its essential features to that ultimately adopted; but it was determined that the application to Parliament should be made on Mr. Green's plans and estimates, and that they should be remodelled by Robert Stephenson after the Act was obtained.

On June 18, 1844, the Newcastle and Darlington line, laid out by George Stephenson as a portion of the great trunk line by the east coast to Scotland, was opened to public traffic, and completed the communication from London to the Tyne. The terminus of this railway was at Gateshead, and consequently the traffic was cut off by the deep chasm of the river from the town of Newcastle, and the railway isolated from the lines already made on the north side, as well as from the proposed continuation of the great trunk to Scotland.

The inconvenience of this was felt most severely. It was found that nearly half a million local passengers frequented the station in the course of a year; and as it was situated on the high ground, they had to descend the steep road, on a declivity of 1 in 8 or 9, to cross the river, and ascend again by a similarly steep hill to the plateau of Newcastle on the other side. The cost of conveyance of passengers and merchandise by coaches and omnibuses across the river reached the enormous amount of £1000 per week.

At this time a considerable agitation took place re

specting the prolongation of the Scotch trunk line from Newcastle to Berwick, for which the route had been surveyed previously by George Stephenson; and this gave a renewed impetus to the question of the bridge. It was seen that as the railway must, if possible, be made to cross the Tyne, it would be highly advantageous to combine both railway and common road crossings in one bridge, the expense of which, though too great to be warranted by the road traffic only, might be very well justified by the addition of that of the railway: and the promoters of the new line wisely saw the advantage that would accrue to their interests if, in conjunction with their works, they could offer to the town the boon of the high-level carriage road crossing, which had been so long desired. They accordingly took up the project of the 'High Level Bridge Company,' and by making it a combined road and railway bridge, incorporated it into their railway survey as an integral part of the scheme.

A rival line to Mr. Stephenson's, promoted by Lord Howick, with Mr. Brunel as the engineer, was, however, started: this was to be worked on the atmospheric plan,* and was intended to cross the Tyne by a bridge considerably to the westward of the town, and at a low level, having gradients of 1 in 50 on each side.

The rival schemes came before Parliament in May 1845; and the High Level Bridge formed one of the most prominent features of the Stephenson line. Mr. Robert Stephenson was examined at some length, and gave a full description of the proposed bridge. Many objections were raised by the opposite party, on the

---

* See the chapter on the Atmospheric Principle of Railway Propulsion.

grounds of the great height of the structure and of the viaducts connecting it with the town—of the risk of the trains running off—of the danger of fire to the houses from the projection of live coke from the engines— of the chance of frightening horses on the carriage road of the bridge by the noise of trains passing over- head ;—and the obstruction the bridge would cause to the ventilation of the town.

These objections were all answered. Mr. Stephenson gave his strong opinion that the plan of the bridge pro- posed was the only one by which the objects aimed at could be properly combined; and after a hard and pro- longed contest, the bill for Mr. Stephenson's line passed, including in it the sanction for the erection of the High Level Bridge across the Tyne.

Mr. Stephenson at once put the design in hand, and the drawings were prepared, under his immediate direction, by his assistant, Mr. Thomas E. Harrison, who afterwards became resident engineer on the line, and is still consult- ing engineer to the larger 'North Eastern' system in which it is now incorporated.

The roadway across the bridge consists of two plat- forms; the upper one carries three lines of railway, while the lower forms the common public road. The approaches of the railway are curves in contrary directions, but those of the public road are in a straight continuation of the line of the bridge. The height of the rails on the upper platform above the low water of the river is 120 feet. The level of the carriage road is about 23 feet lower than that of the railway.

The bridge stands about 90 yards to the west of the

old bridge. The river at this spot is 515 feet wide at high water; but as the bridge has to cover also the sloping shores on each side, its whole length is 1,372 feet. There are six large openings, each of 125 feet span in the clear, stretching over the river and the flat portions of the banks, the slopes being covered by smaller abutment arches up to the high level on either side.

The piers, some of which, from foundation to summit, are as much as 146 feet high, are built of a hard and durable sandstone, obtained from quarries in the neighbouring coal formation. They are founded on piles, the spaces between which are filled up with concrete. Many of the piles are 40 feet long, and all are driven through the hard sand and gravel forming the bed of the river, till they reach the solid rock below. The piles are 13 inches square, and are placed 4 feet apart from centre to centre. The greatest weight that can come upon each of them is 70 tons, supposing none to be carried by the intervening spaces of concrete. This is a very heavy load, which could only be warranted by the goodness of the strata into which the piles are driven. In moderately compact clay it is usual to consider the maximum bearing power of a pile to be about 12 tons; in hard clay about 25 tons; but in gravel, of which the bed of the Tyne consists, 70 or 80 tons are often allowed; indeed, many engineers consider the bearing power in such strata only limited by the resistance of the fibre of the timber. In this case also, the feet of the piles rest on the solid rock, which puts all doubt at rest as to their stability. Mr. Stephenson, however, with his characteristic desire to satisfy his mind thoroughly on the point, tested one of the piles by laying on it a load of 150 tons, which

was allowed to remain for several days; but on its removal no settlement whatever had taken place.

The masonry of the piers commences about 2 feet
below low-water level. The lower portions in the
stream are provided with cutwaters. The foundation
surface of each pier is about 76½ feet by 22½; the
section of the tall shaft of the pier is about 46 feet by
14 feet, lightened by an arched opening 12 feet wide.

But the most important part of the work is the iron
superstructure, and it will be interesting to consider the
motives which led Mr. Stephenson to a decision as to the
nature of the structure by which the openings were to be
spanned. It will be seen by reference to the historical
notice in Chapter II., that in 1845, when this work was
designed, the science of iron bridge construction was
only partially developed. The experiments for the
Britannia Bridge, which had ultimately the effect
of bringing wrought-iron girders into use, had but just
commenced, and the only kinds of iron bridges then
adopted for large spans were—the suspension bridge—
the compound trussed girder—the cast-iron arch—and the
bowstring girder. We may therefore conceive Mr.
Stephenson considering the applicability of each of these
systems in turn. The suspension bridge would have
answered well enough for the common roadway, but it
was inapplicable to the railway from its want of rigidity.
Mr. Stephenson had indeed, just at this time, investigated carefully the possibility of its application to the
Britannia Bridge, and decided against its fitness for railway
purposes. The compound girder of cast-iron, trussed
with wrought-iron bars, had been used by Mr. Stephenson

somewhat extensively, and was at this time being applied by him to the Chester Bridge of 100 feet span;
but he probably shrunk from extending a construction
yet scarcely tried to dimensions so much larger, and to a
situation so much more perilous than anything previously
encountered.

The cast-iron arch must at first sight have recommended itself strongly for adoption. Its principles were
thoroughly known; its strength and stability were indu-
bitable; and it would have made by far the handsomest
structure in an architectural point of view. But to Mr.
Stephenson's far-seeing and eminently cautious professional
judgment, objections revealed themselves which he did
not feel himself able satisfactorily to overcome. Arches
involve outward thrust at their extremities, and to resist
this thrust, so as to keep the whole structure in perfect
equilibrium, great stability in the piers and abutments is
absolutely essential. Now this quality Mr. Stephenson
did not see his way to insuring. The piers were of
great height, and economy demanded all possible saving
in their bulk; so that they would stand up from
the depth below as long slender legs, on the top of
which it would be highly injudicious to allow any
considerable oblique strain to fall; and though the
arches on each side of any pier might theoretically
be supposed to counteract each other's thrust, and
to throw the resultant strain vertically down the body
of the pier, Mr. Stephenson's experience told him this
could not in practice be relied on. But more than
this, he anticipated difficulty with the foundations of
the piers; he knew by his borings he should meet
with treacherous strata; and by calculation of the

weight each pier would have to support, he found that
his bearing piles must sustain a very heavy and unusual load. Under these circumstances he considered
a slight settlement of the piers as a contingency quite
possible, and which he could not with certainty avoid
by the utmost skill and care. And as such a settlement would have endangered in a serious degree the
equilibrium of any arches resting upon the piers, he
deemed it prudent to give up the idea of using the arch
system of construction, to which, it is well known, in
suitable cases he had a strong leaning.

There only remained therefore the bowstring girder,
a form which fortunately combined all the requisite conditions. It was simple in principle, strong and stable,
well understood, and entirely free from the objections
to the arch, inasmuch as, like all other girders, it
was self-equilibrated, gave nothing but vertical weight
upon the piers, and would allow of a slight settlement in
them without serious danger. Moreover, the form of this
girder was eminently adapted for the double roadway, the
top of the arch being at a suitable level for the stiff platform of the rails, while the horizontal tie, or string of
the bow, defined the lighter carriage road. Thus Mr.
Robert Stephenson's comprehensive and acute practical
reasoning enabled him to justify, and indeed to reproduce,
the design which had originally occurred to his father.

He had already built, on the London and Birmingham
Railway some years before, handsome bridges of the
same construction, 50 feet span, which, although they
had only one roadway to carry, had a double horizontal
bar, above and below, as if they carried two ; and it
is not improbable that this feature (introduced there

chiefly for stiffness) may have suggested the peculiar applicability of the form of girder to the purpose of carrying a double road.

Whether, if the High Level Bridge had been designed ten years later, Mr. Stephenson would have adopted the bowstring arch in preference to wrought-iron girders, it is difficult to say : the latter would certainly have been lighter and cheaper, but it would be difficult to find any form of girder, even with all our modern knowledge, that would make so appropriate, so substantial, and at the same time so handsome a bridge as that actually built. It was a mistake in the architectural design to put a pier in the centre instead of a space, but probably this was determined by engineering considerations.

Each span or bay is crossed by four main girders, the chief feature of each girder consisting of a cast-iron arch or bow, the ends of which are connected together, and the thrust taken, by a wrought-iron tension rod or tie.

The rise of the cast-iron arch is 17 feet 6 inches, or a little less than one-seventh of the span. It is made in five segments, strongly bolted and accurately fitted together ; the depth is 3 feet 6 inches at the crown ; the section is that of a double-flanged girder, having 133 square inches area of metal in the two outer girders, and 189 inches in the inner ones, which have more weight to support. The tension ties, or strings of the bow, consist of flat wrought-iron bars, 7 inches by 1 inch, each outer girder being tied by four of these, and each internal girder by eight. The girders are all strongly braced together with diagonal frames.

The railway platform is placed above the top of the

arches, a series of pillars being carried up throughout
the spandrels, so as to support entablature beams lying
horizontally above them the whole length of the bridge.
Cast-iron cross-bearers rest on these, extending in one
length over the four main girders; these again support
longitudinal timber joists, on which a flooring of double
diagonal planking, jointed and tongued with hoop iron, and
well caulked with pitch, is laid; the three lines of rail are
fastened down to the planking in the ordinary way.

The lower roadway for common traffic is hung from the
arches by wrought-iron suspending rods, bearing longitudinal beams similar to those of the top platform, and
upon these cross-bearers rest, carrying a double planked
roadway in like manner. This is paved with wood-blocks
set in pitch, and covered with sand and gravel.

The four girders are so placed as to leave space for a
carriage-way between the two inner ones, with footways
between these and the outer girders on each side. The
carriage-way is a little over 20 feet wide, and the two
footways are 6 feet each The total width of the bridge
from outside to outside is about 46 feet.

Fig. 7 shows an elevation, and fig 8 a transverse section of one of the spans of the bridge, from which a
general idea of its structure may be obtained. The
general elevation is given, on a small scale, in plate at
p. 71 of this volume.

Provision is made for the expansion and contraction of
the iron superstructure by fixing the girders firmly to the
first, middle, and fifth piers, and making them free to move
on the second and fourth, as well as on both the abutments. There are no rollers, but the bearings have surfaces fitted for sliding on each other. The motion caused
by a variation of temperature of 32 degrees was found

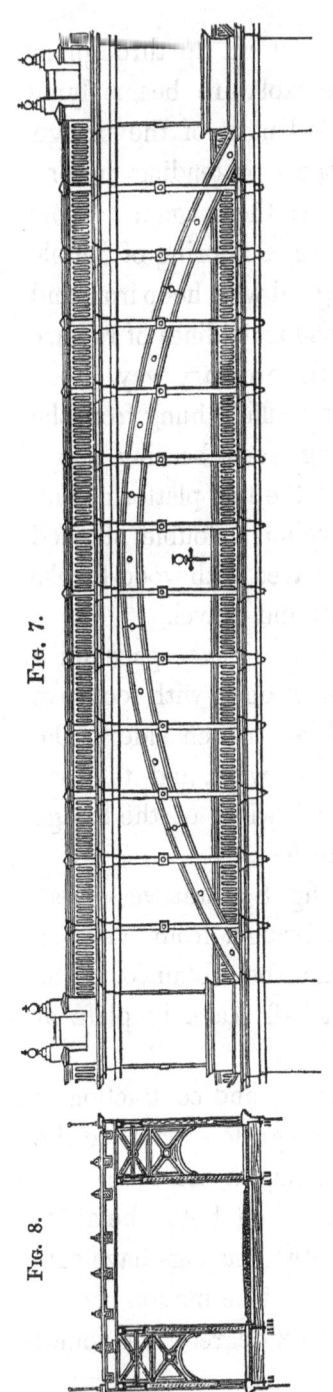

Fig. 7.

Fig. 8.

hy experiment to be 0·153 of an inch for each span.

The land abutments are founded upon a bed of strong clay, which underlies the sand and gravel, no piles being used. They are built of stone similar to the piers, and are of handsome design, carrying the roadways over the slopes on each side by masonry arches of solid construction. Designs were made for ornamental entrances to the bridge, in keeping with the decorative features of the structure; but as the authorities of the railway and the town grudged the few hundreds of pounds necessary for their erection, the work is left incomplete.

The contract for the bridge was let on August 17, 1846, and the work was commenced in October; but many difficulties occurred in driving the piles which considerably retarded the progress of the work; and among others the peculiar effect of ebb and flow during this operation was thought by Mr. Stephenson worthy of special notice in an engineering point of view, as being not generally observed. During flood tide the sand became so hard as to resist almost entirely the utmost efforts of driving, while at ebb it was quite loose, and in no way hindered the operation. It was found necessary to abandon the driving on many occasions during the entrance of the tide.

Another difficulty arose from the quicksands beneath the foundations. Although the piles were driven to the rock the water found its way up, baffling the attempts to fill in between them; this, however, was ultimately remedied by using a concrete made of broken stone and Roman cement, which was continually thrown in till the bottom was found to be secure.

The piles were driven by Nasmyth's steam pile-driver, this being one of the first cases in which it was used; and by its quick action the driving was effected in much less time than by the ordinary means. The ram of the engine weighed a ton and a half, and had a fall of 2 feet 9 inches. It was worked incessantly night and day, driving at the rate of sixty or seventy strokes per minute; and in several instances the pile-heads burst into flame, and burnt fiercely under this rapid action of the ram. It was found by an experiment that after a pile had been driven with the ordinary machine as far as it would go, the application of the steam driver would force it down 15 feet farther.

The coffer-dams for the piers were formed of double rows of piles, filled in with clay puddle; when they were removed, the piles were not drawn, but were cut off level with the bed of the river by a circular saw, the lower parts being left to protect the foundations as well as to avoid disturbing the ground by their extraction.

The ironwork of the superstructure was manufactured by Messrs. Hawkes, Crawshay and Co., of Newcastle; but before any of it was made, Mr. Stephenson instituted a large series of experiments upon different kinds and different mixtures of cast iron, with the view of ascertaining what description of metal would be most advantageous for the purposes of the bridge. These experiments were made with great care, and Mr. Stephenson considered them the most extensive as well as the most accurate series then existing. They are printed in full detail in the Report of the Royal Commission on the Application of Iron to Railway Structures, 1849. They led Mr. Stephenson to the following conclusions :—

1. That hot blast iron was less certain in its results than cold blast.

2. That mixtures of cold blast iron were more uniform than those of hot blast.

3. That mixtures of hot and cold blast iron together gave the best results.

4. That simple samples did not run so solid as mixtures.

5. That simple samples sometimes ran too hard and sometimes too soft for practical purposes.

As far as the construction of the bridge was concerned, the result was that the iron for the principal parts, namely the arch ribs and transverse girders, was selected of the following mixture :—

Ystalifera anthracite	(No. 3)	. .	40 cwt.
Resdale, hot blast	„	. .	40 „
Crawshay (Welsh), cold blast	(No. 1)	. .	40 „
Blaenavon, cold blast	„	. .	30 „
Coalbrook Dale, cold blast	„	. .	30 „
Selected scrap . . .	.	. .	30 „

The first heavy casting was made in February 1847. Each arch was temporarily erected at the manufacturers'

Painted by J W Carmichael 1846

High Level Bridge, Newcastle-upon-Tyne.

Engraved by H Adlard

works and tested before removal, and all the detached parts received a separate test previously to their final trial.

The total quantity of masonry in the bridge is 686,000 cubic feet; the weight of ironwork is 5,050 tons. The cost of the entire work, including that of the temporary bridge, was £243,000.

The iron superstructure was erected on centres supported by scaffolding from below, which likewise answered the purpose of carrying the temporary roadways at different stages of the work. Each bay was divided by a timber pier, leaving a clear opening on each side of about 53 feet; the parts on which each segment of the cast-iron arch rested being strutted from the sides. The segments were lowered to their places by a large traverser or 'Goliah,' running on a tramway about 3 feet below the rail level.

So much importance was attached to getting the railway traffic across the ravine as soon as possible, that it was thought worth while to erect alongside the scaffolding a temporary bridge of timber, which anticipated by a year the opening of the main structure.

The bridge was examined and passed by the Government Railway Inspector on August 13, 1849, and was formally opened by the Queen in the month following.

W. P.

CHAPTER V.

AFFAIRS, PUBLIC AND PRIVATE, DURING THE CONSTRUCTION OF THE CHESTER AND HOLYHEAD RAILWAY.

(ÆTAT. 42–47.)

Newcastle and Berwick Line—The High Level Bridge—Trent Valley Line—Leeds and Bradford Line—Italian Trip in 1845—Norwegian Trip in 1846—Norwegian Liberality—Irish Famine—Lord George Bentinck's Proposal to subsidise Irish Railway Companies—Robert Stephenson, George Hudson, and Mr. Laing consulted—Lord George Bentinck's Speech in the House of Commons—Election to the Council and Vice-Presidency of the Institute of Civil Engineers—Narrow Escape on the Chester and Holyhead Railway—Death of George Stephenson—Relations between Father and Son—Elected Fellow of Royal Society—Grand Banquet at Newcastle—Summary of his Railways—High Level Bridge opened by the Queen—Robert Stephenson declines the Honour of Knighthood—'Nene Valley Drainage and Navigation Improvement Commissioners'—Appointed Engineer with Sir John Rennie to the 'Norfolk Estuary Company'—Consulted by the Town Council of Liverpool as to the best Means of supplying Liverpool with Water—Grand Central Station at Newcastle opened by the Queen—Royal Border Bridge opened by the Queen — Statistics relating to Royal Border Viaduct—Robert Stephenson desirous of Rest.

SIMULTANEOUSLY with the progress of the tubular bridges on the Chester and Holyhead Line, Robert had upon his hands other important works—the Newcastle and Berwick Line, the Trent Valley Line, and many other railways; the High Level Bridge uniting the iron roads on the north and south banks of the Tyne,—and the Royal Border Bridge, spanning the Tweed, and forming a

link between the railway systems of England and
Scotland!

The session of 1845 saw the act passed by which the
promoters of the present Newcastle and Berwick Line
were empowered to carry the Great Northern Line still
farther north. It is needless here to recount minutely
the opposition and defeat of Brunel and Lord Howick,
who (the atmospheric mania being then at its height)
opposed the adoption of the locomotive system on that
important route between the Tyne and the Tweed. The
same session also granted permission for the construction
of the High Level Bridge, a scheme which had been
under discussion during the four previous years.

While the foundations of the High Level Bridge were
being formed, the works on the Newcastle and Berwick
and the Trent Valley Lines were pushed on vigorously.
The viaduct over the Tweed was also under progress.
By July 1, 1846, Robert Stephenson, aided by Mr. T. L.
Gooch, had the Leeds and Bradford Line ready for public
use. On June 26, 1847, the Trent Valley Railway (on
which line Mr. Bidder and Mr. Thomas Longridge Gooch
were co-engineers with Robert Stephenson) was opened
for public traffic. In the following month the Newcastle
and Berwick Line was regularly used for the conveyance
of goods and passengers. There remained only to finish
the great bridges over the Tyne and Tweed, and so
complete an unbroken chain from the capital of Scotland
to the metropolis of Great Britain. Other minor lines
were also under construction at the same time.

But full as his hands were of home work, Robert
Stephenson found time to superintend railway operations in foreign countries. In the summer of 1845,

he joined the Committee formed for carrying out the late
Prince Consort's suggestion for a Grand Industrial Exhibition; and he offered the Committee a loan of £1,000
for preliminary expenses.   In the autumn he visited Italy.
In the long vacation of the following year, severely worn
by the harass of committee-room contests and continued
application to the concerns of his various undertakings,
he broke away from business for a trip of pleasure in
Norway.   Mr. Bidder was his companion; but no sooner
had the two friends become accustomed to the change of
having nothing but pleasure to think about, than the
Norwegian government consulted them on the policy of
uniting Christiania and the Mïosen Lake by a railway.
So impressed was Robert Stephenson with the wisdom of
the proposition, that he offered to defray half the expense
of surveying the line of country.   In that year, however,
the scheme did not proceed beyond consultation.   The
affair was held over for four years, when (in 1850)
Robert Stephenson, having had further negotiations with
the authorities of the state, sent out English engineers
who made the requisite survey, to cover the cost of which
he contributed, as a loan, the sum of £800.   The line at
length mapped out, the works were commenced and
carried on with spirit, Robert Stephenson being retained
as Engineer-in-chief.   The autumn of each of the years
1851, 1852, and 1854 saw Robert Stephenson in Norway superintending the operations.   In 1859 he went
there to receive the congratulations and thanks of the
country on the completion of the enterprise, returning in
his death-sickness to the coast of England.   On his
death, his executors made a demand for his professional
services on the contractors, who forthwith paid the fee,

hardly earned and justly due. On seeking reimburse-
ment, however, from the Norwegian government, the
contractors were informed that the pecuniary remuneration was not 'in the bond.' Norway had already paid
the engineer with the cross of St. Olaff. The grateful
country also repaid the loan of £800, held for several
years *without interest.*

Other work also came upon Robert Stephenson in
1846. Famine had raised the sufferings of the Irish poor
to a point unprecedented—even in Ireland. For once
the humane could calculate with certainty on benevolent
cooperation from the most selfish. In every quarter the
question was heard—how can our fellow-countrymen be
saved from starvation? Lord George Bentinck was earnest
in urging government to subsidise Irish Railway Companies
with funds, so that the crisis might be tided over by
stimulating the demand for labour. Ireland, argued
Lord George Bentinck, had food enough for her poor,
but the poor had no money to give in exchange for it.
If every man in Ireland willing and able to work could
only find employment, there would be an end of the
exceptional wretchedness. There were grounds for this
view of the case. Irish railway projectors sent representation after representation to Lord George Bentinck,
that though they had obtained their acts of parliament,
and in some instances had embarked large sums on
works, they found themselves suddenly brought to a
stand-still by the impossibility of raising funds in the
disorganised state of Irish commerce. It was even
urged, that bridges and other works on the eve of completion had been stopped under circumstances that caused
enormous loss to Irish speculators. Before stirring in

Parliament, Lord George Bentinck consulted Mr. George
Hudson, Robert Stephenson, and Mr. Laing. They were
strongly in favour of subsidising the Irish Railway Companies. With conscientious anxiety to give the country
nothing but sound advice, Robert Stephenson despatched
Mr. Samuel Bidder to Ireland to examine the state of
public works.

Mr. Samuel Bidder writes:—

Before Mr. Stephenson would give any opinion on the subject,
he wished to ascertain the facts as to the amount of work said to
have been already executed. Having received instructions and
plans of the lines from Mr. Stephenson, I made the best of my
way to the several points where it was reported that bridges had
been nearly built, and cuttings half finished; but in almost
every instance I found that not a brick had been laid, a sod cut,
or one shilling expended, and I so reported to Mr. Stephenson.
. . . . I was also requested by Mr. Stephenson to notice,
as I passed through the country, the kind of work that was being
done for the employment of the poor under what was called
the Government Staff. Large gangs I found employed in what
I could call by no other name than the total destruction of the
highways. For the purpose of easing the gradients of the hills
they were removing the crust from the roads, which had taken
years to consolidate; and this was carried on in such a manner
as to endanger the life of every traveller. Many coaches were
upset, and I don't believe a single road was ever improved by
the work. It would have been much better to have fed the
people and saved the roads.

Notwithstanding the misrepresentations made to Lord
George Bentinck, the great fact still remained. Famine
was mowing down the destitute Irish by thousands. It
was clear that the statements of Irish speculators ought
not to be accepted without enquiry; but at the same time,
it was evident that 'labour' was the grand remedy for the
evil, and that while Ireland had need of railways, she had

a vast army of workmen, who were either unemployed or had been set to useless or hurtful tasks.

Under these circumstances, Lord George Bentinck, on February 4, 1847, moved—' That leave be given to bring in a bill for the prompt and profitable employment of the people by the encouragement of railroads in Ireland.' In his speech, introducing the motion, Lord George observes :—

How many men can you, by your scheme, find employment for? We know by experience—at least I know by information from Mr. Stephenson, the engineer of the line—that the London and Birmingham Railway employed 100 men per mile in its construction for four consecutive years. The London and Birmingham Line, however, was one far more expensive in its works than the Irish lines, of which the outside average cost is estimated at £16,000 per mile. The estimate of Mr. Stephenson is, that, taking one line with another throughout Ireland, to execute the whole of them would require the services of sixty men per mile for four consecutive years. Sixty men per mile for 1,500 miles would give constant employment for four consecutive years to 90,000 men on the earthworks and line alone ; but it is estimated that the employment given to quarrymen, artificers, and others, not actually engaged upon the line of road, would occupy six men per mile for the whole number of miles under construction. This would give 9,000 men more; to which is to be added—that which experience teaches us is the fact—that when a new railway passes through a country, the new fences to be made, the fields to be squared, the new drains and water-courses to be cut, and the new roads to be constructed, also occupy at least six men per mile, which will give 9,000 men more, making altogether a total number of 108,000 men. But there are other miscellaneous employments to which the expenditure of so large a sum of money necessarily gives rise, and it is thought to be putting the number very low when we estimate the able-bodied men required to be employed at high wages, in order to accomplish 1,500 miles of railway in Ireland, at 110,000, representing with their families 550,000 persons.

*   *   *   *   *   *

But by this proposition we must expect not only to be able to give subsistence to 550,000 persons, but we seek to provide also for the comforts of these poor people in the course of their employment. We have not forgotten the interests of the labourers; following out the recommendation of the report of the Railway Labourers' Committee, we have inserted in our bill clauses obliging the companies to see that their contractors pay the wages of the labourers once a week, and that in hard cash.

*   *   *   *   *   *

But this is not the only point in which we consider the interests of the railway labourer, and this suggestion comes from my honourable friend Mr. Stephenson. It is that the companies shall be required, on the demand of the Railway Commissioners, to construct decent and suitable dwellings for the labourers before they commence their works. Nothing can be stronger than the language used in the report which lies on the table on this point. It states truly, that it is in vain to think of improving the morals of the people except you begin by improving their social condition. In practice, however, it has been found that railway labourers have been generally crowded into dwellings and put in places not fit for pigs. Some may think this measure an interference with free trade in the construction of railways; but I understand, from all the best contractors in the kingdom, that it is cheaper to them in the end to consult the convenience and comforts of their labourers. Experience teaches that if a man is uncomfortable at home, he will go to the public-house; and that where labourers cannot be comfortably provided for, and have no opportunities for bringing their wives with them, they will get tired of their work, and desert it altogether.

This was Robert Stephenson's scheme for Irish Relief. His care and labour, however, proved all in vain. On the third night of discussion, Lord George Bentinck's bill was ' put off for six months ' by the votes of a full House.

In the following summer Robert Stephenson himself entered the House of Commons, as member for Whitby.

Having acted as Member of the Council of the Institution

of Civil Engineers from 1845 to the close of 1847, Robert Stephenson became a Vice-President of that learned Society in 1848, which office he held till the close of 1855, when he took possession of the Presidential chair for the years 1856 and 1857.

In the year 1848 he met with an accident, which may be briefly noticed. On August 20, accompanied by his friend Mr. Lee, he entered a first-class railway carriage in the Conway Station ; when, as soon as they had taken their seats, the carriage was pushed across the down line in order that it might be put in position on another pair of rails. Before the carriage was quite clear of the down line, the Chester ' express,' coming up at full speed, caught its projecting angle and threw it about a yard from the rails, smashing the glass doors, wheels, and frame-work. Immediately after the concussion, Robert Stephenson was seen descending the steps. In another half minute, he was observed to fall on his back, the shock of the collision having deprived him of power to walk. Mr. Lee, who had seen the approach of the ' express,' and had provided for the accident by pressing his back and feet firmly to the padded sides of the carriage, escaped unhurt. The next morning Robert Stephenson was well enough to proceed to Chester.

Just eight days before this narrow escape, death gave Robert Stephenson the severest blow he had experienced since he put his wife in her grave at Hampstead. On August 12, 1848, George Stephenson died suddenly, after a brief illness, in the sixty-seventh year of his age. His death was altogether unexpected ; only seven months before,* he had married a young wife, and even on the

---

* George Stephenson married his third wife on January 11, 1848.

day of his death was looking forward to many years of happiness.

The closing years of George Stephenson's life were bright with success. He had seen the locomotive brought from the rudeness and imperfection of the Killingworth Engines to an efficiency that has not yet been greatly surpassed, and he had seen his son rise to be the leader of his profession.

Robert Stephenson had always been a devoted son. In all his quarrels and contests, George Stephenson was sure of his sympathy and support. But of all the modes by which the latter endeavoured to add to his father's happiness, the most beautiful was his habit of uniting him in the glory of his achievements. When the designs for the High Level Bridge and the Tubular Bridges were under discussion in the elder Stephenson's presence, the son always spoke of them as ' *our* works at the Straits,' or, ' *our* bridge over the Tyne.' In this graceful manner did Robert repay the love of the father, who inscribed the name of ' his boy ' on the first plans of the Stockton and Darlington Line. On George Stephenson's death, the portrait of Robert, painted by Lucas, passed (in accordance with the understanding between the subscribers for it and the elder Stephenson, to whom it was presented for life) into the possession of the Newcastle Literary and Philosophical Society. The picture was painted in 1845.

George Stephenson left great wealth behind him, but even if Robert had loved his father much less than he did, an accession of wealth would have been no consolation to him for the bereavement. By his own exertions, he had acquired as much wealth as he desired. Without a child, unmarried, and resolved never to marry again, he had no ambition to be very rich.

A distinction highly prized by men of science was conferred on the inventor of the Tubular Bridge in 1849. He was elected a Fellow of the Royal Society, on the council of which learned society he subsequently sat.

At this period Robert Stephenson turned his attention to other fields of engineering. In 1848 he was consulted by the River Nene Improvement Committee, who were interested in the improvement of the river between Peterborough and the county boundary near Wisbeach. From that date up to the time of his death he maintained a professional connection with the ' Nene Valley Drainage and Navigation Improvement Commissioners.' *

In the following year he and Sir John Rennie were appointed Engineers-in-chief of the Norfolk Estuary Company, at which time he reported on the Norfolk Estuary Scheme, and was examined thereon by the Admiralty Commissioners and both Houses of Parliament. The matters which for years remained in protracted ' dispute between the Norfolk Estuary Company and the Eau Brink Commissioners, relating to the mode of executing a cut and works below Lynn, under the Norfolk Estuary Acts,' would little interest the general reader, but they involved important interests, and for years gave the engineers concerned in them much anxiety and labour.

At the close of 1849 the Liverpool Town Council

---

* River Nene. Report of Robert Stephenson, Esq., M.P.; G. P. Bidder, Esq., and Geo. Rob. Stephenson, Esq., on the Improvement of the River Nene, pursuant to resolutions passed at a Public Meeting held at Wisbeach, Sept. 10, 1857, with a Prefatory Letter from Mr. Robert Stephenson, 1858.

consulted Robert Stephenson as to the best means for securing an adequate supply of water to the town of Liverpool. The conclusions of the engineer on this subject may be found in a report sent in to the Water-Committee of the town council on March 28, 1850.

On Tuesday, July 30, 1850, Robert Stephenson was entertained at a grand banquet by four hundred gentlemen on the platform of the new railway station, in the Forth, at Newcastle. Three excellent cartoons (the work of Newcastle artists—Mr. John Storey, Mr. John Gibson, and Mr. R. S. Scott), representing the engineer's grandest works — the High Level Bridge, the Menai Tubular Bridge, and the Royal Border Viaduct, ornamented the enclosed space. The chair was occupied by the Hon. H. T. Liddell, the son of George Stephenson's first patron, Lord Ravensworth.

Proposing on this occasion the health of Mr. Thomas Harrison, Robert Stephenson said :—

No one felt more intensely than he did the value of the assistance which he had derived from those who had been associated with him for some years past. If they would read the biographies of all their old distinguished engineers, they would be struck with the excessive detail into which they had been drawn ; when intelligence was not so widely diffused as at present, an engineer like Smeaton or Brindley had not only to conceive the design, but had to invent the machine and carry out every detail of the conception ; but since then a change had taken place, and no change was more complete. The principal engineer now had only to say let this be done, and it was speedily accomplished, such was the immense capital, and such the ample resources of mind which were immediately brought into play. He had himself, within the last ten or twelve years, done little more than exercise a general superintendence ; and there were many other

persons in that room to whom the works referred to by the
chairman ought to be almost entirely attributed. He had had
little or nothing to do with many of them beyond giving his
name, and exercising a gentle control in some of the principal
works. In that particular district, especially, he had been
most fortunate in being associated with Mr. Thomas Harrison.
Beyond drawing the outline, he (Mr. Stephenson) had no right
to claim any credit for the works above where they now sat. Upon
Mr. Harrison the whole responsibility of their execution had
fallen, and he believed they had been executed without a single
flaw.

The completion of Robert Stephenson's Northumbrian
works had another and still more memorable celebration.

When the Queen opened the High Level Bridge in
1849, she was so cordially received by her Northumbrian
subjects, that she readily consented to repeat her visit to
Newcastle and Berwick in the following year, at the opening of the Grand Central Station at the former, and the
completion of the 'Royal Border Viaduct' at the latter
place. As soon as it was known that Her Majesty would
visit Castle Howard, the Earl of Carlisle's seat in Yorkshire,
in order that she might conveniently be present at both
towns during the same day, and reach Holyrood Palace
before the evening, great exertions were made on the
banks of the Tyne and Tweed to give her an appropriate
welcome, on August 29, 1850.

In the evening of that day a dinner took place at the
Assembly-rooms, the mayor occupying the chair, and
Robert Stephenson sitting by his side. The engineer had
just declined the honour of knighthood, which Her
Majesty had expressed her readiness to confer upon him.
He had reached a point of life and fame when such
rank could afford him neither pleasure nor profit. If

his wife had been still alive he might have decided
otherwise.

The opening of the Royal Border Bridge is an important point in the life of Robert Stephenson. It was the
last of the great works with which he enriched his native
land—the grand conclusion of many years of toil that
scarcely knew relaxation. The preceding ten years had
added much to his glory, but they also had made cruel
inroads upon his physical power. From 1840 to 1850,
he had never known a day free from grave care. No sooner
was one stupendous undertaking brought to a close,
than others rose to take its place. For the greater portion
of that time he had under his care many distinct affairs,
any one of which would have overtasked the powers of a
man of ordinary capacity. It had been a long fierce
struggle with difficulties. The atmospheric contest, the
battle of the gauges, the tubular bridges, the catastrophe
at Chester, were features of the retrospect.

It was now time that he should rest in some degree
from labour.

He formed a plan of withdrawing gradually from
professional turmoil; and if he only in part carried out
this resolution, the fact is not to be attributed to change
of intention, but to the determination of others to make
use of him to the last. So far, however, did he adhere to
his purpose, that he never again entered upon any important undertaking at home. He was always ready to give
advice to his professional brethren, and indicate the course
to be adopted by projectors who sought his counsel.
But he never again became personally responsible for
the success of an important work in Great Britain. And
if foreign powers had allowed him to follow his own

inclinations, he would most probably till his dying day have remained content with the fame which he had won.   But such complete freedom from responsibility was not permitted.   The Norwegian government solicited him to make good his promise to give them a railway ; Egypt begged him to introduce the new locomotion to the inhabitants of the desert ; and from the other side of the Atlantic a petition came that he would throw a tubular bridge over the St. Lawrence.

CHAPTER VI.

ROBERT STEPHENSON AS POLITICIAN AND MEMBER OF THE HOUSE OF COMMONS.

(ÆTAT. 44-55.)

George Stephenson's Political Opinions and Sympathies — Robert Stephenson's Toryism — '*Little* Lord John! — Opinions on Popular Education — Robert Stephenson M.P. for Whitby in Yorkshire — 'One of the Impenetrables' — Speech in the House of Commons on the proposed Site for the Great Exhibition of 1851 — Discussion on the Army Estimates, June 19, 1856 — First Speech against the Suez Canal — Second Speech against the Suez Canal — Speech on 'The State of the Serpentine' — Popularity in the House of Commons — Letter to Admiral Moorsom on Crimean Mismanagement — Reason for declining the Invitation of the Newcastle Conservatives — Dislike of Party Strife — Testimonial to Sir William Hayter.

IN his biography of George Stephenson, Mr. Smiles has rightly observed that the political opinions of the elder Stephenson 'were at best of a very undefined sort.' To think closely and logically on matters unconnected with his mechanical enterprises was not his habit. Like most men of imperfect education, he was guided by emotion, rather than reason, in the consideration of subjects that lay apart from his daily avocations. But his *sympathies* were strongly conservative. To those who conceive of the ambitious workman, fighting a long fight against adverse circumstances, as a person necessarily smarting under a sense of social injustice, it may be matter of surprise that the preeminently successful workman of

Great Britain, at a time when the operative classes were very generally at variance with the classes above them, was throughout his career well disposed towards existing institutions. The lessons of life had taught him to look to the aristocracy with loyal attachment. His first friends in the Newcastle neighbourhood, who had taken him by the hand, were amongst the leading persons of Northumbria. Lord Ravensworth had been his patron. Mr. Brandling had distinguished him with a support widely different from the countenance ordinarily bestowed by a wealthy and well-descended proprietor on an ingenious mechanic. His early patron Mr. Losh, also, was placed high above the common rank of manufacturers by culture, attainments, and public services. From such men the workman had received encouragement, when encouragement was most needed ; and in after-life he was warmly attached to those superior classes from which his earliest friends came.

Robert Stephenson heartily concurred with his father's political sympathies; but he was a politician in whom intellectual conviction went hand in hand with sentiment. He prided himself on being 'a Tory, the term conservative being by no means strong enough to express his abhorrence of innovation. It would only raise a smile to enumerate the articles of his political creed, which resembled that of Colonel Sibthorp more closely than that of Lord Derby. Give the government of the country for a month into the hands of Cobden, Bright, and Roebuck, and (he maintained) the national preeminence of Great Britain would be lost for ever. The bare mention of Earl Russell's name invariably ruffled his temper. Some of his views were the prejudices of a previous generation. He even maintained that the superficial culture obtained by workmen at reading-clubs and lecture-rooms injured them as artisans. A really good mechanic ought to be bent on achieving *manual* perfection, throwing all his strength of body and soul into the special task assigned him; and such earnestness in comparatively uninteresting labour the captain of workmen deemed incompatible with general mental culture. At a dinner of the Royal Society's Club, he made a party of *savans* open their eyes with astonishment by exclaiming, 'It is all nonsense Lord John preaching and preaching education to the working classes. What the artisan wants is special education for his own particular specialty. And the more he leaves everything else alone the better.' This speech was made in the last year of his life (February 21, 1859), at a time when he had displayed rare munificence in the cause of popular education, and when he was meditating still greater contributions to the same cause. So strangely does an earnest and amiable man's practice differ sometimes from his theory!

In the summer of 1847, at the same crisis when his father declined to stand as a candidate for the representation of South Shields, Robert Stephenson was invited by the electors of Whitby, in Yorkshire, to become their member of parliament. The invitation was unanimous. All shades of political opinion were merged in a common desire to pay a well-merited compliment to the man who had been tested by years of arduous service. At any time this expression of confidence would have been agreeable to him; but coming to him when his reputation was under the cloud temporarily cast over it by the failure of the Dee Bridge, it made a deep and permanent impression on his heart.

On Tuesday, July 27, 1847, Robert went to Whitby, and was received at the railway station by his committee and principal supporters. On the following Friday, his election to the seat took place without opposition.

For Whitby Robert Stephenson continued to sit in parliament till his death. In 1852 he was opposed by the Hon. Edmund Phipps, when a majority in his favour of about two to one warned politicians not to renew the attempt. Unquestionably his Tory supporters had no reason to regret placing confidence in him. The atmosphere of the House strengthened rather than weakened his political convictions. Protectionist to the marrow, he disdained to relinquish his belief in Protective principles, even though persistence in them brought him in direct collision with experience. To his dying day he argued warmly in favour of the great commercial fallacy, but it was remarked by his most intimate friends that it was the one solitary subject about which in his last years he would in discussion lose his temper ; and that exceptional irritability appeared to them a sign that his confidence in arguments rejected by the rest of the world had been gradually impaired. But beyond irritability he made no sign of penitence, and as far as avowal went, he was in 1859 as much an 'impenetrable' as he was in 1852, when he voted in the memorable minority of 53 against the equally memorable majority of 468.

As a member of parliament Robert Stephenson voted steadily with his party, but he abstained from taking part in the debates, unless the Commons stood in need of his professional information or judgment.

On July 4, 1850, when Colonel Sibthorp made his speech and motion against the proposal for the Great

Exhibition of 1851, Robert Stephenson, who, as an un-compromising Tory, was not without points of sympathy with the Colonel, spoke out boldly against the motion of the member for Lincoln. The question was altogether distinct from party politics, and the engineer who was himself a staunch promoter of the enterprise and a member of the Building Committee, pleaded the cause of the Exhibition with good effect, in a speech which, as a maiden speech from a member of high celebrity, made a most favourable impression on the House.

The next occasion of his addressing the House was on March 25, 1852, when he spoke briefly but emphatically in favour of the London (Watford) Spring Water Company Bill. On the 17th of the following June, he again spoke a few words. In the debate on the second reading of the Metropolitan Burials Bill, Lord Ebrington drew attention to the defective drainage of the metropolis, and more especially to difficulties encountered by the contractors for the Victoria Sewer. When his lordship resumed his seat, Robert Stephenson rose and said :—

The whole of the evils complained of in the construction and increased expense of the Victoria Sewer was in consequence of the course pursued by the Woods and Forests. He went to the Chief Commissioner, and endeavoured to induce him to permit the sewer to be discharged into the Thames within the limits of the Crown property; but that request was peremptorily refused. He saw no reason why Crown property should be exempted, or, at least, why it should stand in the way of the best system of drainage for the metropolis. But he received a peremptory command that the sewer should not be discharged within the limits of the Crown property. This refusal made a large difference in the expenditure, and, worse than that, the foundations having been bad in the new line, a great portion of the sewer was now in a very dangerous state. Whether the responsibility of this rested on the late Chief Commissioner (Lord Seymour) or not, he would not undertake to say.

For the next three sessions Robert Stephenson was content to be a silent member of the House. On June 19, 1856, he took part in the discussion on the Army Estimates, making a few practical remarks on the subject of Ordnance Maps.

He must (he said) admit that each scale had its peculiar advantages, yet at the same time he must contend that for practical references the one-inch map was the best of all. The habit which unfortunately prevailed in this country of continually changing the scale occasioned a profitless expenditure of public money and endless confusion. The great *desideratum* was a map of Great Britain on a uniform scale. In Ireland the scale had been altered once or twice, and the result was most deplorable. He had frequently consulted the six-inch maps in that country, but invariably found them worse than useless for engineering purposes. It gave him nearly as much trouble to gather information from them as to realise it on the ground. What the engineers employed in various parts of the United Kingdom had for years been endeavouring to obtain from Parliament was an assurance that the one-inch map would be completed before any other piece of surveying was taken in hand. If it were to be continually interrupted, as it had heretofore been, fifty or sixty years would elapse before it was perfected. At present there was no uniform map of England on which engineers could depend. In France and other countries through which he had travelled on professional business, he had never found any difficulty in procuring lucid and accurate maps; but such could not be said for England, though the English had probably spent on their surveys ten times as much as any other people in Europe. Over and over again the engineers of Great Britain had given the government to understand that maps to the six, the twelve, or the twenty-five-inch scale were alike useless to them, but the Ordnance turned a deaf ear to all their protestations, and went on constructing maps which the engineers did not want, and would not consult.

The next parliamentary speech made by Robert Stephenson was his most important address to the Commons. It was important both to the public at large and to himself, because it confirmed the general dislike to the proposal for a Suez Canal, and it pledged his professional judgment to the unsoundness of the scheme.

On July 17, 1857, Mr. Griffiths asked the House ' Whether in their deliberate opinion it be conducive to the honour or the interests of this country that we should manifest and avow the existence of a jealous hostility on our part to the project of a ship canal through the Isthmus of Suez ? ' In reply, after recapitulating the political objections to the proposed canal, and upon them condemning the scheme as ' one which no Englishman with his eyes open would think it desirable to encourage,' Lord Palmerston observed : —

As regards the engineering difficulties, I am aware there is nothing which money and skill cannot overcome, except to stop the tides of the ocean, and to make rivers run up to their sources. But I take leave to affirm, upon pretty good authority, that this plan cannot be accomplished, except at an expense which would preclude its being a remunerative undertaking ; and I therefore think I am not much out of the way in stating this to be one of the bubble schemes which are often set on foot to induce English capitalists to embark their money upon enterprises which, in the end, will only leave them poorer, whomever else they may make richer.

Robert Stephenson then rose, and said : —

He would not venture to enter on the political bearings of the subject with respect to the other powers of Europe, but would confine himself merely to the engineering capabilities of the scheme. He had travelled, partly on foot, over the country to which the project applied, and had watched with great interest the progress that had been made by various parties in examining

the question. He had first investigated the subject in 1847 in
conjunction with M. Paulin Talabot, a French engineer, and M.
de Negrelli, an Austrian engineer. At the suggestion of Linant
Bey, a French engineer, who had been upwards of twenty years
resident in Egypt, and feeling how important was the establishment, if possible, of a communication between the Red Sea and
the Mediterranean, he had qualified himself to form an opinion
on the subject. It had been received on the authority of an
investigation of the levels taken by the French engineers during
the invasion of Egypt about 1800, that, as stated by the ancient
writers, there was a difference between the levels of the Mediterranean and the Red Sea of something like thirty-two feet. It
was suggested at that time that the old canal might be opened
out again, and that a current might be established between the
Mediterranean and the Red Sea of from two to three miles an
hour, which velocity of water would not impede the communication between the Mediterranean and the Red Sea, as
steam tugs might be employed, and the canal might at the same
time be kept perfectly open, as the scouring power would be
adequate to maintain a clear channel. He went into this
scheme under the belief that that difference in the level did
actually exist. The examination was made by himself and the
gentlemen with whom he was associated in 1847. They had
not any idea, at that time, that if there was no difference of level
it would be practicable for a canal to be made in the first instance,
or that it could be maintained afterwards. After investigation,
however, it was found that, instead of a difference of thirty-two
feet there was no difference of level whatever, at the period of
low water, although for a period of fifty years the world had
been under the impression, from the published statements and
levellings of M. Lepère, that a difference of thirty-two feet
existed ; and whilst it was supposed to exist, it was believed by
professional men that a canal might be maintained, or that, as it
was called, a new Bosphorus might be formed between the Red
Sea and the Mediterranean. But when the difference of level
was found to be *nil*, the engineers with whom he was associated
abandoned the project altogether, and he believed justly ; and
one of them (M. P. Talabot) made an adverse report, which was
published in the 'Revue des deux Mondes' of May 1855.

Since then he had travelled over the Isthmus to Suez, and over other parts of the Desert, and had investigated the feasibility of making a free communication between the two seas, on the supposition that they were upon the same level—as, for instance, from the Nile. He might, however, say, without entering into professional detail, that he had arrived at the conclusion that it was—he would not say absurd, because engineers whose opinions he respected had been to the spot since, and had declared the thing to be possible—at all events, if feasible (as the First Lord of the Treasury had said, money would overcome every difficulty) yet, commercially speaking, he frankly declared it to be an impracticable scheme. What its political import might be he could not say, but as an engineer he would pronounce it to be an undesirable scheme, in a commercial point of view, and that the railway (now nearly completed) would, as far as concerned India and postal arrangements, be more expeditious, more certain, and more economical than even if there were this new Bosphorus between the Red Sea and the Mediterranean.

In the following year (June 1, 1858), Mr. Roebuck made and Mr. Milner Gibson seconded a motion to the effect that, ' in the opinion of the House, the influence of this country ought not to be used in order to induce the Sultan to withhold his assent to the project of making a canal across the Isthmus of Suez.' The question having been thus again raised in the legislative assembly, Robert Stephenson in the course of the debate said :—

As the member for Sheffield (Mr. Roebuck) had made special reference to himself in the course of his observations, he begged to occupy the attention of the House for a few minutes. The learned member had told them that it was very desirable to facilitate the intercourse between one portion of the globe and another. No one doubted that, but the speech had not satisfied them that this canal would accomplish that object. It assumed that it would, but he (Mr. Stephenson) believed that it would not do so. On the contrary, even supposing its construction to be physically possible, which he, for one, denied, he was prepared to show that the engineering difficulties would render the scheme impossible. In attempting to prove the feasibility of the project the member for Sheffield had quoted many authorities, but he had omitted to refer to the opinions of the three gentlemen, one from Austria, another from Paris, and himself of England, who first investigated the subject in 1847. They examined the physical features of the country, and deliberated over the matter in the most cautious manner, basing their observations upon the erroneous supposition that it would be possible to establish an artificial Bosphorus between the Red Sea and the Mediterranean, such as existed naturally between the Black Sea and the Mediterranean. They proceeded upon the assumption that the French levels were accurate, which showed a difference of thirty feet between the one sea and the other, by means of which a constant current would be maintained through the channel, supposing it to be made sufficient to keep the harbour of Pelusium free from the mud which was deposited from the Nile. Instead of there being a difference of thirty feet in the height, however, it turned out that the two seas were on a dead level, and that no current whatever could be established. The member for Sheffield therefore was almost guilty of a misapplication of terms when he spoke of a 'canal,' because if this channel were cut, and the water let into it, it would not be a canal, but a ditch. The speaker had quoted the late Mr. Rendel as a supporter of the scheme which was now advocated; but he (Mr. Stephenson) could say positively that Mr. Rendel did not support the scheme as now proposed, and he might mention as a proof of this that he did not sign the report. Mr. Rendel must have been known to every honourable gentleman in that House, and his authority on such matters was very great. Mr. McClean, another high authority in matters of this description, also denied the feasibility of the project. As far as the English engineers were concerned, he believed they all agreed with him to a man. With respect to the difficulties in the way of carrying out the scheme, he would only point to the difficulty of cutting a canal through a desert, with no fruits, no fresh water to be found within that space. He had travelled on foot the whole distance, at least over all the dry land, and consequently felt justified in what he stated. He did not desire to enter upon the political part of the question, but he could assert that as far as the transit of passengers and mails was concerned, the proposed scheme would be productive of no saving of time in our intercourse with the East; for, while they could be conveyed from Alexandria to Suez by railway in eight hours, it would require, even if the most perfect canal possible were constructed, at least double that time for vessels going to India to pass through it, for vessels must coal either at Alexandria or Suez. It was said that they had nothing to do with the physical difficulties of the scheme, but he thought the House had something to do with them, or at least he had. If he had sat silent it would be said he had acquiesced in the motion, and had tacitly admitted that the Suez Canal was a feasible project, whereas his opinion was that if it were attempted at all—which he hoped it would not be, or, at least, he trusted it would not be with English money—it would prove an abortive scheme, ruinous to its constructors.

In the July of the following session Robert Stephenson spoke several times on the Metropolis Local Management Act Amendment Bill. The last time that he appears to have addressed the House was on August 11, 1859, when he made a speech of considerable length in the debate on the *State of the Serpentine*.

It is therefore seen that Robert Stephenson was by no means the silent member that he has been represented. On committees and at divisions he was always at his post when wanted; and he spoke, not eloquently, but fearlessly, impartially, and with authority, whenever he felt he could impart valuable information to *the House*. When he wished to address his constituents, he went to Whitby for the purpose; and on such occasions he met with a cordial reception, even from political opponents. His speeches in the House on the Suez Canal gained for him much ill-feeling. He was accused by many of being prejudiced against a grand project; a few even went so far as to insinuate that his opposition sprung from interested motives. The great body of his fellow-countrymen, however, felt grateful to him for braving anger and detraction, in order that he might save the public from what he deemed an unsound speculation.

In the House of Commons Robert Stephenson was extremely popular. Men of all parties were glad to hold friendly intercourse with him; and he thoroughly enjoyed their generous recognition of his claims as an eminently successful man of action. In the smoking-room, representatives of all shades of political opinion clustered round him for chat, for notwithstanding his extreme Tory principles, he numbered amongst his personal friends the most important of his political antagonists. How little Robert Stephenson permitted politics to influence him in affairs of private friendship is well illustrated by the following hearty letter to his old friend Admiral Moorsom, in which he playfully speaks of himself as 'a good old Tory.'

34 Gloucester Square, Hyde Park:
Feb. 1, 1855.

DEAR ADMIRAL MOORSOM,—Your letter arrived at my office a week after I had left for a short tour in France. Otherwise I should have been very glad to have had a quiet discussion with you on the subject of the attack on Sebastopol. Although it is a subject on which I could not have brought much knowledge to bear, yet with your assistance I think we might have got up a very good *pro and con.* I should like a good old Tory (without reference to conscience) have supported all that has been done, and you, I take it for granted, would have propounded something new. To be serious for a moment, I am quite incapable of giving an opinion upon the mode of attack, but on the management of the commissariat it is impossible to form but one opinion. It is distressing, execrable, and contemptible. There is not one redeeming feature in the whole thing. I have been in France for the last month, and everywhere *quiet* Frenchmen stand aghast at our management. They say little, because they have become half Englishmen for the time, and deplore the sad casualties as much as we do ourselves. But still their countenances indicate a great deal more than their tongues are willing to give utterance to.

The government has been turned out. I don't blame them half so much as I do the heads in command. They are the true delinquents, or rather the true incapables. The truth is, we have lost the art of war amongst the officers. The men have proved themselves equal to everything within the reach of human beings, but the officers are utterly devoid of science. There has not been, as far as I can judge, a single indication of a mental effort. Inkerman is creditable in point of science to the Russians, discreditable to our chiefs in command, although glorious to our regimental performances.

How true what one of the French generals said, ' The charge was magnificent, but it was not war.' A compliment was never before draped in so much satire. As an Englishman I sincerely wish he had never said it, because it is so just, so laconic, that it will live through all time to our disgrace. If I had been in stronger health I think I should have been off in the Titania, and I am not sure that I shall hesitate much longer. I long to be at Balaklava to get the stores away to the camp. I believe I could be useful there as long as I felt myself beyond the range of the artillery. When nigh them I am persuaded I should run away and disgrace my country.

Yours faithfully,

Admiral Moorsom.　　　　　　　　　　　ROBERT STEPHENSON.

It should be borne in mind that, Tory as he was, Robert Stephenson supported the Earl of Aberdeen's Government and voted against his party in the division on Mr. Roebuck's motion in January 1855.

Never did political ardour manifest itself with greater gentleness in any man than it did in Robert Stephenson. Sincere as he was fervent, he had a lively sense of the respect due to honest opinions. In 1852, there was a movement in Newcastle to elect him for that important constituency; but he declined the proffered services of the Tyneside Conservatives, on grounds which par-

liamentary candidates are wont to esteem too lightly.
' I should,' observed Robert Stephenson, writing to his old
friend Mr. Michael Longridge (June 3, 1852), 'have to
ask many of my friends to vote for me at the expense
of their political convictions, which I could not stoop
to do.'

For party strife he had a strong dislike.   In 1848, he
did his utmost (short of sacrifice of principle) to avoid
wounding for a second time the feelings of M. de Lesseps
and his partizans, but when he found himself compelled
either to repeat his condemnation of that which he
deemed an unwise project, or to countenance it by
silence, he did not hesitate to take the manly course.
' You see,' he wrote to Mr. Thomas Longridge Gooch,
' I have been pitching into my dear friend Lesseps again
about the Suez Canal.   I believe they feel that the thing
is *squashed* for awhile, Roebuck was determined to bring
the matter again before the House.   . . .   I tried to
get him to withdraw the motion, as I knew any fresh
discussion upon it could only engender bad feeling on the
part of the French.   He was perfectly resolute, and
would not listen to any course but the one he had proposed for himself.   I had therefore no alternative but to
repeat what I had formerly said, and to stop, as far as
I could, the English people from spending the money on
an abortive scheme.'

One of Robert Stephenson's last acts deserves to be
recorded in a chapter that surveys his career in the
House of Commons.   When Sir William Hayter resigned
the office of ' whip,' the duties of which he had discharged
for many years with great efficiency, and with good
results to the liberal party, his political friends presented

him with a testimonial. Hearing of the subscription for the testimonial, Robert Stephenson went to the committee and asked leave to be a subscriber. In reply to the request, it was represented to him that as the testimonial was to be presented in recognition of services to the *liberal party*, the committee could not accept of a contribution from a *Conservative*. The decision of the committee was unquestionably prudent, and in accordance with good taste ; but it was far from satisfying Robert Stephenson, who represented in forcible terms that he wished to subscribe, not as a political personage, but as one of Sir William's oldest and warmest private friends. So urgent was he in the matter, that Lord Mulgrave brought the question again before the committee, when Robert Stephenson was permitted to pay Sir William Hayter as flattering a compliment as friend can pay to friend.

CHAPTER VII.

ROBERT STEPHENSON IN LONDON SOCIETY.

(ÆTAT. 47–55.)

The Year of the Great Exhibition—In the Park—An Impostor imposed upon—No. 34 Gloucester Square—'The Sunday Lunches'—Works of Art—Philosophical Apparatus—Demeanour in Society—'The Chief' in Great George Street—Robert Stephenson and 'the Profession'—Stories of Robert Stephenson's Generosity—'The Westminster Review' on Robert Stephenson—Cab-drivers and their Payment—Zenith of Robert Stephenson's Prosperity—His part in the Great Exhibition of 1851—Crack-brained Projectors—Aquatic Amusements—The 'House without a Knocker'—Alexandria and Cairo Railway—Victoria (St. Lawrence) Viaduct—Mr. Samuel Bidder's Reminiscences—Grand Banquet at Montreal to Robert Stephenson—His Speech on the Occasion—Connection with Mr. Alexander Ross—The St. Lawrence Viaduct completed, and inspected by Robert Stephenson's Deputies.

THE year 1851 was the period when Robert Stephenson may be regarded as at the fullness of prosperity. There were few names more honoured, no man more generally popular. On questions either immediately or remotely connected with engineering, to state Robert Stephenson's opinion was in general society to terminate discussion. In the House of Commons and the clubs he was always welcome—his sociable disposition rousing sympathy by means of a fine presence, a countenance singularly frank, an unaffected *bonhomie*, and that pleasant richness of voice that impresses the hearer with an idea of intellectual and moral excellence. He was to be seen frequently in the parks, where as he took his riding exercise he was pointed out to visitors from the country as one of the most notable of existing ' lions.'

Following the fashion of the day, Robert Stephenson, between 1851 and his death, used at times to wear beard and moustaches; at other times his lips and chin were shaved.

Of his martial appearance, at the times when his beard and moustache were at their fullest, a good story is told. Ten or twelve years since, in one of the streets of Soho, there was a pretender who undertook to tell persons' characters by an infallible method. He employed a mechanical contrivance composed of a crystal globe and a framework, placed in the centre of the chief table in his reception-room; the crystal globe being suspended from the frame in such a manner that it oscillated like the pendulum of a clock. Of course any disturbance to the frame or the table affected the vibrations of the glass ball. The ' wise man's ' visitors were separately made to take up a position against the table, leaning or lounging upon it, and by observations of the movement of the globe and of the images upon it, while the table was subjected to such pressure, the seer professed to be able to tell the mental and moral characteristics of those who consulted him. The imposture made a sensation. The knave was honoured with visits from great ladies; and the gossip of the ladies induced men of scientific reputation and high political influence to call on the quack. Amongst many others, Robert Stephenson went to look at the charlatan. In an ante-room leading into the chamber

devoted to the wizard and his globe, he was requested to
write his name and address in a book. Instead of com-
plying with the request, he entered in the register the
name and address of a friend, of whom there was small
chance that the character-reader would have any knowledge ; that done, he passed on to the inner room, reclined
against the table, and exchanged a few sentences with his
entertainer. He took his leave, with the understanding
that his ' character,' duly written out, would be sent in
the course of the next day to his residence. According
to promise, the ' character ' arrived at the given address :
' Frankness and decision are your leading characteristics.
Of timidity and caution you are altogether ignorant.
Bold, fearless, dashing, reckless, you would make an admirable cavalry officer, but it is clear that you are
utterly devoid of mechanical talent.' Such was the pur-
port of ' the character.' Robert Stephenson's imposing
height and moustache had completely imposed on the
impostor.

At 34 Gloucester Square, Robert Stephenson entertained
his friends with liberal hospitality. Few private enter-
tainments in London were more pleasant than his ' Sunday
lunches,' at which many chiefs of literature and science
were in the habit of meeting. Baden Powell, Sir R.
Murchison, Sharpe (the Egyptian scholar), M. Bonomi,
Captain Pim, Sir James and Lady Prior, Mr. Lough (the
sculptor), Dr. Mayo, Dr. John Percy, Brunel, and other
not less eminent persons came to Gloucester Square for
these receptions.

Robert Stephenson left Cambridge Square for Gloucester Square somewhere about the November of 1847.
The purchase of the lease of his last residence (together

with sums expended on alterations and repairs) involved an outlay of nearly £10,000.

Amongst the more important works of art contained in the house at the time of his death, were—

1. A full length, life-size portrait of George Stephenson, painted by Lucas.

2. 'The Evening Gun,' by Danby. This picture was exhibited at the Manchester Exhibition. After Robert Stephenson's death, Mr. George Robert Stephenson presented it to Mr. Bidder.

3. 'The Twins,' by Landseer. On this picture Robert Stephenson set great value. An opulent gentleman, breakfasting with him one morning, offered him £5,000 for it. 'But,' said the owner, repeating the circumstance of the offer to his friends, 'he stood a worse chance of getting it by setting so high a value on it, as I knew him to be an excellent judge.'

4. 'Killingworth Colliery,' by Lucas.

5. 'The Stepping Stones,' by Lucas. This picture, painted at Robert Stephenson's order, represents a girl carrying a child over a stream in Wales. The Britannia Bridge is seen in the distance.

6. Portraits of engineers and others, grouped in consultation with Robert Stephenson about the Britannia Bridge, by Lucas.

7. Railway arch at Newcastle-on-Tyne, by Richardson.

In sculpture Robert Stephenson manifested his taste by purchasing Power's 'Fisher Boy,' the companion of the 'Greek Slave,' in the Exhibition of 1851.

The drawing-rooms of 34 Gloucester Square were so liberally stocked with works of curious contrivance, and philosophical toys, that they had almost the appearance of a museum. Singularly constructed clocks, electric instruments, and improved microscopes, by Smith and Beck, and Pillischer were arranged on all sides.

Robert Stephenson's cabinet of microscopic specimens

was most elaborate and extensive; and any contribution to it was an attention he always cordially acknowledged. He had long desired to have some specimens of North American coal, and also of coal brought by him from South America, prepared for the microscope; but no one had been able to reduce the mineral substances to the necessary degree of thinness. After months of ineffectual labour, however, Mr. Stockman had the good fortune to get some specimens sufficiently thin for the purpose. On receiving them Robert Stephenson was greatly delighted, and gave expression to his satisfaction by presenting Mr. Stockman with a costly microscope.

In society Stephenson was a charming companion. A ready talker, he was also a courteous listener. Never presuming on his reputation and position, he encouraged perfect freedom of discussion, and even on questions of engineering, he would hear patiently, and answer with respect, the views of his opponents. He was never guilty of dogmatism towards the young, or superciliousness to the timid. It would be difficult to imagine a man more considerate of the feelings of others. No description of his demeanour in the society of men would be complete which did not contain the word 'jolly.' He was the embodiment of joviality, without the faintest touch of boisterous awkwardness. 'I never in all my life knew a more clubable man than Robert Stephenson: it is impossible for a more clubable man than Robert Stephenson to exist,' is the emphatic testimony of Dr. John Percy. But merely one-half of the man's social capabilities were known to those who saw him only in the society of men. His courteous bearing to ladies possessed the style of ancient chivalry.

In his own profession, Robert Stephenson's popularity
was even greater than his popularity in general society,
or with the public.   Perhaps no man of such eminence
has ever had less of jealousy and detraction embittering
success.   In Great George Street he exercised a sort of
feudal sway over a numerous body of engineers, who had
served either under him or his father.   By them he was
always designated the ' chief,' and whenever any of them
undertook a new work, ' the chief' was the first person
to whom all the particulars of the enterprise were confided.
No one was afraid of him.   His amiable and admirable
power of making men of the most uncongenial tempers
' pull together' gained him the affection of men whose
dispositions were far from gentle or conciliatory.   ' Robert
Stephenson *made me*!' one gentleman says, gratefully recalling his ' chief's' influence ; ' I never was able to get on
with other men, till Robert Stephenson took me in hand.'
Men like Brunel, with whom he was continually at issue
on public questions, never for a moment regarded him as
' an enemy ;' whilst most of them found him a generous
and unselfish friend.

His liberality to professional subordinates in respect of
money, was equal to his generosity towards them in
respect of all that concerned their reputation.   Every
engineer working for him was secure of twice the payment he could get from any other employer, and was
moreover certain that the merit of his labour would
not be appropriated by the general-in-command.   Two
anecdotes, taken from many others, will display this side
of Robert Stephenson's character.   At a time when the
Stanhope and Tyne embarrassment made money a great
object to him, he was engaged as engineer to a branch

line in one of the midland counties, with a fee of £6,000,
the directors permitting him to find an engineer to
execute the works, so long as he visited the line from
time to time, supervising the operations, advising on all
important points, and personally bearing all the responsibility. Under these circumstances, the engineer-in-chief
sent down Mr. A. to execute the work, on the understanding that they should divide the fee, taking £3,000
each. Before a third of the line had been put down,
Mr. A. received an offer of more lucrative employment,
and asked Robert Stephenson if he might resign his post,
or find a *locum tenens.* On receiving permission to
adopt the latter course, Mr. A. made the work over to a
friend (Mr. B.), on agreement that they should divide the
moiety of the original fee, taking £1,500 each. On this
arrangement the line was completed ; when, the time for
settlement having arrived, Robert Stephenson sent Mr. A.
a cheque for £4,000 instead of £3,000. Mr. A. called on
Robert Stephenson to say that the sum was £1,000 in
excess of the amount due; and the latter said, 'I am
aware that by our agreement you can demand only
£3,000; but as half your fee goes to Mr. B., whose
services *we* have benefited by, I think we three engineers
had better divide the £6,000 equally, share and share
alike: men should never try to eat each other up.' This
story was communicated by the gentleman designated as
Mr. A., but he requested that the name of himself and his
locum tenens should not be published.

Another case is given by an engineer, who desires
(out of regard to the feelings of other persons) that no
names should be mentioned in its recital. On one of his
more important railways, Robert Stephenson, as engineer-

in-chief, and the members of the engineer's staff, were
engaged by the directors on fixed salaries, commencing at
the time of their engagements, and terminating on the
completion of the line.   During construction, monetary
derangements brought the works to a stand-still for many
months, and threatened ruin to the company.   The consequence was that for the best part of a year the subordinate engineers had nothing to attend to on the line,
and were able to take employment elsewhere.   When
the question was raised whether the salaries should be
paid for the period extending over a complete cessation
of work, Robert Stephenson, who felt that it would be
hard for an impoverished company to pay heavy salaries
to an unemployed staff, and that it would be no less
for the subordinate engineers (none of whom were rich
men) to relinquish their claims, devoted his own salary to
satisfying the demands of his assistants.   In point of fact,
he stepped in between the company and the assistant
engineers, and removed the grounds of dispute by paying his staff out of his own pocket.

Hundreds of stories like these might be told. 'If,'
says one of the first iron-masters of the kingdom, ' Robert
Stephenson grew rich himself, he made others richer also.
No one ever had dealings with him, without being the
richer for them.'   But while he distributed freely with
one hand, he was far from clutching eagerly, and holding
fast with the other.   Money always seemed to him, business man though he was, a comparatively unimportant
consideration.   Of course, in some years, his professional
earnings were very great.   It would be difficult to state
them within a few hundreds, or even thousands, for the
engagements of an engineer in large practice are very

complicated, his work being extending over years, and frequently the day of payment being postponed till long after the completion of the work. Possibly Robert Stephenson earned, during the years of most active railway speculation, £30,000 a year ; but whether he earned more or less than this sum, there is no doubt that he could have greatly increased it, if his palm had itched for fees. Latterly he declined to give his evidence before committees, or to advise professionally, unless he received more than ordinary payment; but this did not arise from greed of gain, so much as from indifference to it. He asked for more in order that he might have less. He wished to get rid of his employers.

In estimating Robert Stephenson's conduct to his professional brethren, it should be remembered that he always took all the responsibility of works on his own shoulders. Whatever mistakes might be made, he always took all the blame of errors to himself, and shielded his assistants from criticism.

He was not without his little whimsicalities ; but they were more remarkable for amiability than eccentricity. The greatest and best paid trafficker of his day in the commodity of locomotion, he could never do otherwise than regard those humble dealers in the same article — namely, cab-proprietors and cab-drivers — as an ill-used class. He always insisted on paying cab fares by a scale of his own. The Hansom was his favourite vehicle, and the driver was always required to drive at a brisk pace, the remuneration being at the rate of a penny a minute, from the moment of hiring ; no fare of course being computed under a shilling. Friends frequently exclaimed against the extravagance of this penny a minute rule,

pointing out to him that he did much harm by a liberality
that made the drivers discontented with arrangements
which, while they defended the public from extortion, any-
how allowed hackney coachmen to get a living. 'The
law, which you do your utmost to make people discontented with, is one that especially considers the poor
traveller,' observed a critic, bringing his arguments to a
conclusion. 'Exactly,' answered Robert Stephenson,
warmly; 'and that's just the reason why rich travellers
should not take advantage of it.'

Robert Stephenson's success was at its fullest in the
years 1850 and 1851. His fame and popularity never
reached a higher point. He had ample leisure for the
enjoyment of general society; and his health, though far
removed from all that a man who had only entered on
middle life might reasonably desire, did not shut him out
from enjoyment. In those years he acted as member of
the Executive Committee and Building Committee for the
Exhibition Building of 1851, and also as one of the Royal
Commissioners * for carrying out that noble undertaking.
The testimony of a fellow-committeeman on the Building

* Reference to the Journals of the
Royal Commission, ascertained that
Robert Stephenson attended meetings of the Executive Committee on
January 8, 11, 25, 1850; meetings of
the Building Committee June 21, 22,
1850; Meetings of the Royal Commission, March 7, April 25, May 2,
9, 16, June 21, July, 1, 11, 15, 16,
in the year 1850; March 5, 29,
April, 22, 28, May, 10, 12, July 5,
August 19, October 13, in the year
1851; March 1, April 2, October 29,
in the year 1852; January 25, 1853;
June 30, 1855; June 20, 1857;

May 14, 1858. In addition to these
attendances, Robert Stephenson was
a constant attendant at the Contract
Committee.

As early as June 1844, Robert
Stephenson was a promoter of the
Industrial Exhibition. When it
was attempted in that year to act on
the late Prince Consort's suggestion
to the Society for Encouragement of
Arts, Robert Stephenson became a
member of the provisional committee, and offered a loan of £1,000
to carry out the undertaking.

Committee, as to Stephenson's services to the Great Exhibition, deserves place. Writing (Oct. 26, 1859), Mr. Donaldson says of Robert Stephenson and the younger Brunel:

Being a member of the Building Committee for the Exhibition Building of 1851, with them, and the then Mr. W. Cubitt, C.E., Sir Charles Barry, and Professor Cockerell, it was impossible for me not to be struck with the ready imagination, the brilliant ideas, and the quick invention of Brunel, and the thorough master of detail, the sagacity of perception, and rapidity of calculation with which Robert Stephenson examined each project presented to the Committee, calculated the effective strength of the parts, and threw out suggestions for some grand and novel features that ought to distinguish an erection so novel in its destination, and which admitted so large a margin for novelty and invention . . . . . . When Robert Stephenson was about to raise the last tube of the Menai Bridge, he attended one of the meetings of the Exhibition Building Committee, and invited the members to go down to witness that interesting operation of a gigantic work. Professor Cockerell was then ill, and other members were otherwise engaged; but Sir C. Barry and myself accepted the invitation, and went down. After the last tube had been floated to between the piers, and had been raised and deposited in the place which, we may hope, it is destined to occupy for centuries to come, there was, as usual, a friendly dinner party, and those present congratulated Robert Stephenson on the complete success of his magnificent conception, and expressed the cordiality with which they participated in the triumph of so remarkable a work of science, which had met almost insuperable difficulties, and had created a new application of a system of constructive combination never hitherto imagined. In reply, he thanked his friends for their sympathy and expressions of friendly regard; but he added, that even the triumph of that day did not recompense him for the days and nights of anxious toil and thought, the cares and anxieties which had attended the work.

Robert Stephenson had no country establishment;

though there were few men who had more need of a
place of retreat from the bustle of London life.   Throughout 'the season' he was, during his last years, literally
persecuted by importunate projectors.   When casual in-
disposition kept the engineer from Great George Street,
and confined him to his house, swarms of talkative,
and for the most part profitless clients intruded on the
privacy of the man whose too pliant temper laid him
open to their annoyance.   An intimate friend and col-
league of Robert Stephenson says that, calling in at
Gloucester Square to consult the master of the house
on urgent business, he found every reception-room occupied by a crowd of persons.   Being much engaged,
and wishing to employ his time with correspondence till
he could have an interview with Robert Stephenson, the
narrator asked the servant to show him into a room
where he could be by himself, and write his letters in
quiet.   'If you want that, sir,' the man answered, 'you
must go upstairs into one of the bedrooms, for every
sitting-room is occupied with gentlemen who insist on
seeing Mr. Stephenson, although they know he is unwell.'
And the caller, acting on the advice, went upstairs and
sate in a bed-room, till he could be admitted into the
library.

Amongst these importunate intruders were amateur
engineers with proposals for improved railway breaks.
They formed a distinct class of inventive monomaniacs,
and were so numerous, that Robert Stephenson adopted
an excellent method to protect himself from them.   Of
every stranger who appeared in the ante-room at Great
George Street, Mr. Sanderson (Robert Stephenson's
brother-in-law and business manager) was directed to

enquire the object of his visit.   Whenever the reply was 'Sir, I have a proposal for a new break, about which I should like to ask Mr. Stephenson's opinion,' the following conversation would ensue:—

Mr. Sanderson. 'Indeed, sir. And what can you do with the break?'

Inventor. 'I can stop a train instantly—instantaneously, sir.'

Mr. Sanderson (with an expression of horror in his countenance). 'Good Heavens, sir! if you did that, you'd kill all the passengers.'

Inventor (suddenly modifying his statement). 'But, sir,— I can stop a train gradually.'

Mr. Sanderson (bringing the discussion to an end). 'So can anybody else.'

After this the inventor seldom cared to prolong the interview.

What the country is to most London men of business, the sea was to Robert Stephenson. To any shooting-box within the four seas he would have been followed by clients and letters; but once afloat he was safe from the intrusion of callers and raps of postmen.

In 1850, his first yacht, the Titania, 100 tons, (or 'the *old* Titania,' as she was subsequently named) was launched. A graceful vessel and a fair sailer, the *old* Titania had won a secure place in her owner's affections, when she was destroyed by fire at Cowes. When Robert Stephenson did not require her services, she was always at the disposal of Mr. George Robert Stephenson, who in the second year of her existence sent a line to the captain, ordering fires to be lighted on board, and the yacht prepared for the reception of himself and family. The first part of the order was carried out with such an excess of zeal, that an overheated flue led to the destruction of the vessel.   When Mr. George Robert Stephenson arrived at Cowes, he found the Titania thoroughly gutted.

Making the best of his way back to London, Mr. George Robert Stephenson hastened to Gloucester Square with intelligence of the accident.   On arriving there, he found his cousin entertaining a party of friends, who had just sate down to dinner.   The soup and fish were still on the table.   'What brings you here?   What's the news?' asked Robert, with his customary cordiality.   'Bad news, Robert!' was the answer, 'I ordered the Titania to be got ready for me, to take me and my family a few days' cruise; and—she's burnt to the water.'   Certainly the announcement was far from pleasant; but Robert thought less of his own loss than his cousin's chagrin.   'Well, well, man—don't be annoyed.   You couldn't help it. Sit down and have your dinner.   We'll talk about it over a glass of wine.'   After receiving an account of all the particulars of the mishap, Robert Stephenson put his hand kindly on his cousin's shoulder, and said, 'Never mind, old boy, we'll have a finer vessel than the *old* Titania, before we are many months older.'

Mr. Scott Russell therefore received instructions to build another yacht; and on June 21st, 1853, the *new* Titania (184 tons) was launched.

Soon after the destruction of the *old* Titania, Robert Stephenson wrote to Admiral Moorsom:—

24 Great George Street:
May 25, 1852.

MY DEAR ADMIRAL MOORSOM,—I am really much obliged to you for your mention of the work on Naval Architecture, for I am not quite daunted by the loss of my very fine vessel, as I am again cogitating with Scott Russell over the lines of another yacht. *I find I can get no peace on land. I am therefore preparing another sea lodging-house.* I find it no easy matter to get rid of a multitude of questions which follow on a tolerably long professional life. Indeed I find that nothing gives me actual freedom from attack, but getting out of the way of the postman. *The sea, therefore, is my only alternative. Ships have no knockers, happily.* I shall read carefully the work to which you have drawn my attention before deciding finally on the details of my new vessel, and in doing this I have no doubt that I shall be able to say that I am indebted to you.

Faithfully yours,
ROBERT STEPHENSON.

Admiral Moorsom.

Not quite three months later Robert Stephenson wrote to the same correspondent : —

24 Great George Street, Westminster:
August 6, 1852.

MY DEAR ADMIRAL MOORSOM,—I fear that I shall not have an opportunity for some time to come to visit Sheerness or Chatham, as I am about to leave for Norway in a few days.

I am quite inclined to think with you that a rectangular box with a proper bow and stern is the most likely form in plan for a huge line of battle ship ; but the question had nearly resolved itself into one of section, and even in this respect I am persuaded the rectangle is not far wide of the mark. The approach to the wedge of Sir W. Simons is clearly more than enough for stability, and it cannot fail to make a ship with armaments on board roll quickly and consequently strain everything to pieces. The rectangular tub will carry the day.

Yours faithfully,
ROBERT STEPHENSON.

The Titania (184 tons burthen, drawing 13 feet of water, 90 feet in length, and 21 feet in breadth of beam), was a yacht for a sailor to criticise with approval.

Somewhat deficient in speed, it had every other good quality. The saloon and sleeping cabins were large, the former being 16 feet by 15 feet, and 8 feet high.

Robert Stephenson's yacht was generally allowed to be the best manned in the squadron. In 1856, when Professor Piazzi Smyth, the Astronomer Royal for Scotland, was sent with very limited means to Teneriffe, to make scientific observations, Robert Stephenson placed the Titania and her crew at the professor's disposal.

It was on board the Titania that ' the best ' of Robert Stephenson was seen to ' best advantage.' So elated was he with the sea-air, that the first twenty-four hours afloat were usually passed in school-boy hilarity. Every Sunday that he spent in his yacht he assembled the whole crew, and read them the service appointed for the day by the Church of England.

His delight was ' to show the sea ' to a novice who could endure the motion of the waves, but was inexperienced in the delights of a nautical trip. To take such an one out for three or four days, and bring him home after ' a good blow,' was a genuine pleasure to him. Here is a glimpse of the man, enjoying himself on deck. The late Mr. Kell, of Gateshead, writes :—

In October 1857, Mr. Robert Stephenson was in his (second) yacht, Titania, with a party of friends who had accompanied him through the Caledonian Canal and along the western coast of Scotland to Holyhead. On the morning of the 16th, they sailed from Holyhead for Kingstown in a stiff breeze from the south-west. There had been a dense fog, and many land birds had been blown out to sea. On the passage, the yacht was boarded by several of them. When Mr. Stephenson came upon deck, the crew had captured some linnets, and having taken them into their fore-cabin, were in pursuit of others. Those which had been captured, and taken below, were sitting

in nooks, and would not feed.　Mr. Stephenson said they would
all die, and ordered them to be set at liberty and the pursuit
discontinued.　Several linnets and a lark afterwards came on
board the boat, hanging at the stern of the yacht, and then
hopped upon deck.　Mr. Stephenson became much interested
in getting the birds to feed, offering them crumbs, which
they would not eat.　They attempted to drink the water
washed on deck, but its saltness was painful to them.　Mr.
Stephenson then tried the experiment of steeping slices of bread
in fresh water, and placing them before his little guests.　To
his great delight, the plan succeeded.　Three of the linnets
became very tame, and had to be driven from their seats on the
ropes.　When the working of the yacht required them to change
their position, Mr. Stephenson always took care that they were
not injured.　The other linnets and the lark could not get over
their terror at the disturbance in shifting the sails, and after
returning to the ship several times, disappeared—no doubt to
perish.　As soon as the yacht entered Kingstown Harbour, the
three linnets flew off, clung to the tangle on the dock-wall for
some minutes, and then with apparently renewed vigour flew
on shore, Mr. Stephenson exclaiming, 'There, you little un-
grateful birds, we have saved your lives and made Irishmen
of you.'

In the autumn of 1850, Robert Stephenson went on a
trip of pleasure to Egypt,—a trip that, like his excursion
to Norway, led to another important engagement.　Whilst
he was at Alexandria, *en route* for England, the then
Viceroy of Egypt, Abbas Pasha, summoned him to Cairo,
for a conference on a scheme for introducing railway
communication into Egypt.　The result of the intercourse
between the Pasha and the engineer, was a commission
from His Highness the late Said Pasha to connect Cairo
and Suez by a line of railway.　In consequence of con-
sular representations the first scheme was abandoned,
and the line between Alexandria and Cairo was decided
on, chiefly from the influence of Mr. Hugh Thurburn.

In 1851, Nubar Bey was sent to London to make the requisite contracts, and on the return of the envoy the works were forthwith commenced and carried on with great spirit, so that the main line was available for public traffic in the opening months of 1856. The break of way, caused by the intersection of the main line by the Nile, between Kaffr Zeyat and Kaffr Lais, which is now spanned by the magnificent Kaffr Zeyat Tubular Bridge, was in the first instance met by the engineer with an enormous steam ferry, consisting of a huge pontoon, which supported a high staging furnished with a railway deck. Constructed so that it could be raised more or less in accordance with the varying depth of the waters of the river, this ponderous vessel received the trains, with all their burden, and conveyed them to the opposite bank.

At the close of 1856, Robert Stephenson, accompanied by Mr. Frederick Richard Lee, R.A., and Mr. Sopwith, visited the railway, and found the ferry at work. In an account of the trip, printed for private circulation, Mr. Sopwith gives the following account of the floating pontoon.

At Kaffr Zeyat the river Nile bends in a horse-shoe form midway or very nearly so, as has been already stated, between Alexandria and Cairo. This horse-shoe bend is about three miles in length, the intervening land being little more than a mile in width. In the middle of this curve, at a point where the river is 1,200 feet in width, is a ferry which conveys the railway carriages and wagons, as well as the passengers, from one side to the other.

Ferry boats, guided by chains, are used in some places in England, and were first adopted on a large scale by the late Mr. J. M. Rendel at Plymouth; but in the present case peculiar difficulties were presented, which required the especial care of

Mr. Stephenson, under whose direction the railway and all its
accompaniments were constructed. Impediments existed which
prevented his adoption of a bridge, and a ferry being the only
alternative, a design was made and the works were executed at
Newcastle. To-day was the first opportunity Mr. Stephenson
enjoyed of seeing the work in operation: and I must here
attempt, however imperfectly, to convey some idea of the
excusable gratification which he experienced, and the extreme
interest and delight with which I viewed the great engineering
work.

The dimensions of the ferry-boat and its superincumbent
framework are—eighty feet in length, the same in height, and
sixty feet in width. It is worked by two steam-engines placed
horizontally, one on either side, and together of about thirty-
horse power, which suffices to take this gigantic frame of
wrought iron, with a moderate railway train and passengers,
across the Nile in six minutes. Its central portion is a huge
parallelogram, supported by iron buttresses from the projecting
sides of the boat. The striking peculiarity of its construction
arises from the necessity which exists to adapt the floor and
rails which receive the carriages to the exact level of the railway
at each side, under the great variation of level to which the
waters of the Nile are subject, and which at times has been
known to be as much as twenty-seven feet. By an ingenious,
yet simple application of screws, this platform is raised or
depressed, as the water of the Nile may require; and this,
together with an exact adjustment of the rails, and the utmost
facility of access to and egress from the framework, are admirably
accomplished. In England, or indeed in Europe generally, or
in North America, no limits would be placed on the inventive
genius of an engineer, by any consideration as to the ability of
workmen for conducting requisite operations; but in Egypt,
where a prejudice exists in favour of employing native, and
consequently unskilled labour, it was necessary to combine, if
possible, superabundant power with the utmost simplicity of
management: and this has been attained so completely that the
Arabs, receiving less than fourpence daily wages (out of which
they have to purchase bread, their only provision), are quite equal
to the nicest adjustments of this ponderous affair . . . . . . To

contrive and execute such a work at Newcastle, and send it
some three thousand miles, was a work which Mr. Stephenson's
knowledge of Egyptian climate and requirements, and his
proficiency as a practical mechanic, enabled him to undertake
with reasonable and well-founded hopes of success. The experience of to-day, however, proved that a mind more sanguine
than his would have been requisite to rely with certainty on the
entire fulfilment of the accuracy which was aimed at. Seldom
does it come within the good fortune of the ablest contrivers,
even at home and amidst all the means and appliances of able
cooperation, to effect all that they purpose to accomplish: and
some slight misgivings—some vague doubts—might readily
accompany the conception of so large a work for so distant a
land. Six months or more have elapsed since its completion,
and the first view of the perfect success and easy management
of this noble ferry seemed to exceed the best anticipations of its
designer, and I could not but participate in some degree in the
laudable pride and pleasure of such a result. A brilliant sun
and unclouded skies imparted great beauty to the scene, and I
trust that the photographic views to be taken by Mr. Lee will
supply in a great measure, the inevitable deficiencies of my
verbal description.

But well as the ferry worked, it was found inadequate
to deal with the rapidly increasing traffic of the new railway; and subsequently a fixed bridge was substituted for
it. This bridge, which is known as the Kaffr Zeyat
Bridge, was designed by Mr. George Robert Stephenson,
after the model of the other Egyptian iron bridges, and
was completed in 1860. Of the tubular structures over
the Nile at Benha, and over the large canal at Birket-el-
Saba, the engineer's terse description is the best account
for the general reader. In his article on 'Iron Bridges'
in the Encyclopædia Britannica, Robert Stephenson
says:—

The principal feature of the Egyptian Railway Bridges is, that *the road* is carried upon the top of the tubes, and not in their interior.

There are two tubular viaducts upon the Egyptian Railway. The larger one crosses the Damietta branch of the Nile at Benha, and the smaller one crosses the Karrineen Canal at Birket-el-Saba (Lake of the Lion). These viaducts unite two old roads, formerly connected by a ferry, and each is contiguous to a vice-regal palace.

In the larger viaduct there are ten spans or openings, the two centre ones comprising one of the largest swing-bridges that has been attempted.

The total length of the swing-beam is 157 feet; it is balanced in the middle of its length on a large central pier. When open to the navigation a clear water-way is left on either side of the central pier of 60 feet. Each half of the beam sustains its own weight as a cantalever, 66 feet long.

The eight remaining spans are 80 feet in the clear, arranged four on each side of the centre portion, and the total length of the viaduct between the abutments is 865 feet.

The piers consist of wrought-iron cylinders, 7 feet in diameter below the level of the Nile, and 5 feet in diameter above that level. They were sunk by a pneumatic process to a depth of 33 feet below the bed of the river, through soil of a peculiarly shifting character, and are filled in with concrete.

There are six of these cylinders in the central pier which supports the swing-bridge; and the adjacent piers on either side of the centre have each four cylinders; each of the remaining piers has two cylinders only. The tops of the cylinders are covered by cast-iron circular plates, which rest entirely upon the concrete, special care being taken to prevent any contact with the cylinders. On these circular plates rest the upper cast-iron plates which connect the piers, and form a seating for the bearing-plates of the beams.

The beams or tubes are 6 feet 6 inches deep, and 6 feet 6 inches wide at the bottom, tapering to 6 feet wide at the top, and they rest at their ends on rollers working between planed surfaces to admit of the motion caused by expansion and contraction.

The tubes carry a single line of way on their tops, the rails

being laid on longitudinal sleepers; and there is also a roadway
4 feet wide on either side, supported by wrought-iron brackets
bolted to the sides of the tube.

These roadways are of corrugated iron, resting on the brackets,
and stiffened by strips of bar-iron placed transversely on the top.

The six cylinders for the central pier are also provided with
cast-iron circular plates, as before described, and surmounted by
a framework of cast iron, uniting the tops of the cylinders, and
serving as the lower tramway for the rolling machinery.

The revolving machinery consists of a turn-table containing
eighteen accurately turned conical rollers, their angle being
determined to the greatest nicety, and corresponding with the
angular surfaces of the tram-plates between which they revolve.

The diameter of this turn-table is 19 feet from centre to
centre of the rollers.

The whole of the rollers, together with the wrought-iron
circular frame to which they are attached, form an independent
system, usually termed the 'live-ring,' held in its position by
the central pivot. The frame of the 'live-ring' is connected
with the rollers by radial spindles, with gun-metal gudgeons at
the periphery and centre. And to prevent any difference in
angular speed between the rollers and centre portion, a very
excellent arrangement is adopted, which consists in a diagonal
strap passing over the centre wheel, and extending to the outer
periphery. This strap is keyed up to any adjustment, in which
it firmly keeps the radial spindles.

The swing-tube is firmly attached to the upper tram-plate by
a system of cast-iron bracket work, and strong bolts and nuts;
forming, in fact, as is most essential at this point, an exceedingly
rigid attachment. The centre pivot is of forged iron, 9 inches
diameter, and turned accurately to fit its bearings. To insure
a firm fixing for the pivot, it is made to pass through the entire
depth of the lower tram-plate into a socket provided for the
purpose, in which position it is firmly keyed. The bridge is
turned with facility by a capstan worked by two men, with
gearing communicating with the large rack surrounding the
lower tram-plate.

To prevent accident to the swing-bridge when open,
'fenders' are placed up and down stream, similar in construc-

tion to the piers of the bridge. At the bearing ends of the swing-bridge arrangements are made for locking it in its position. These consist of fixed inclined planes attached to the under-surface of the bearing ends of the tube, and corresponding wedges which slide on the piers, which are made to recede and advance, by means of a screw turned by gear-work.

In the Birket-el-Saba Viaduct, the swing portion forms spans on each side of 43 feet, and the fixed portion consists of two spans of 70 feet each. In other respects the viaducts are precisely similar.

Both were commenced in May 1853, and completed for traffic in October 1855.*

In the year 1852, Mr. Alexander Ross, who had care-fully examined the natural features of the St. Lawrence

---

* The engineer's staff in Egypt was an able one. Divided at about mid-way by the Nile, at Kaffr Zeyat, the Alexandria and Cairo Railway had two separate corps of engineers,— one for the district north, the other for the district south of the Nile. Acting as resident engineer, and su-pervisor of both divisions, was the late Mr. M. A. Borthwick. Under Mr. Borthwick were Mr. Henry J. Rouse (having the ordinary control of the northern half of the line), and Mr. Swinbourne (having the entire southern half). At his head-quarters in Alexandria, Mr. Henry J. Rouse was assisted by Mr. Pringle; and at Cairo, where Mr. Swinbourne had his head-quarters, the principal sub-assistant engineer was Mr. Duff. Belonging to Mr. Rouse's corps were also the following engineers—Mr. Anger and Mr. Bidder, Jun. (stationed between Kengis Osman and Damanhour), Mr. J. H. Stanton and Mr. Joseph Harri-son (appointed to the part between Damanhour and the Nile). Belong-ing to Mr. Swinbourne's division were—Mr. Fowler and Mr. Vaughan stationed at Benha; Mr. Rushton and Mr. Hardcastle employed on the line between Cairo and the Nile. Be-sides these engineers, were the sur-veyors, Mr. Graham and Mr. Preston (with Mr. Cheffins, Jun., as assistant), and the architect of the stations, Mr. Edward Baines.

The task of designing the Benha and Birket-el-Saba bridges, and of supervising their manufacture in England, was placed by Robert Stephenson in the hands of his cousin Mr. George Robert Stephenson, who, aided by his managing assistant, the late Mr. George Barling, completed them most successfully.

Every portion was manufactured in England, and tested before being sent from the country. The immense machinery for opening and closing the swing-bridge, was put together and tried, previous to its transmission to Egypt.

at Montreal, and had even prepared a plate iron tubular
bridge for Mr. John Young (whom Canada has greatly to
thank for the bridge of which she is so justly proud),
consulted the inventor of tubular structures. Robert
Stephenson gave the project his most earnest consideration,
visited Canada, produced an elaborate design for the
bridge, and as engineer-in-chief directed the operations
that resulted in the ' Victoria (St. Lawrence) Bridge.'

In February 1853, the survey for the bridge, on the
site it now occupies, was entered upon. In the following
August, Robert Stephenson crossed the Atlantic, personally inspected the shores of the St. Lawrence, had inter-
views with the Directors of the Grand Trunk Railway of
Canada, reconsidered his designs, settled the terms of his
contract with the company, and with a half promise to
visit Canada again in the course of the following year,
returned to England, leaving Mr. Alexander Ross on the
spot to carry out the plans agreed upon.

Of this Canadian trip, Mr. Samuel Bidder gives the
following particulars :—

I had the pleasure of accompanying Mr. Stephenson to the
North American continent, and will relate one incident which
you may think worth recording. We travelled from Portland
into the State of Maine to Montreal, a distance of 300 miles,
by a special train drawn by a Bogie engine, the first known
engine of this kind having been built by Mr. Stephenson himself,
which, it appears, was copied by the Americans as being the
most suitable for their railways.

Mr. Stephenson had never had an opportunity of seeing what
kind of road such an engine could travel over with safety before
the present occasion.

The railway through the White Mountain district had only
been opened a few days, and we were told that many of the
embankments had sunk 5 or 6 feet, but they (the directors),

'guessed we should get along pretty safe;' and sure enough we found the embankments looking more like the hollows between the crests of the waves in the Atlantic, than anything else I can compare them to. Mr. Stephenson and myself stood on the platform of the cars, and had to hold on to the rails by both hands, as hard as possible, to prevent ourselves from being thrown off, for over this road we went, and round curves not more than 200 yards radius, at a speed of from 15 to 20 miles an hour, and I shall never forget the expression of Mr. Stephenson's countenance during the journey. On our arrival at the boundary line, which divides Canada from the States, Mr. Stephenson jumped off, and examined the engine, and said it was the most dangerous ride he had ever had in his life, but that he was not sorry he had had an opportunity of proving and witnessing what this extraordinary machine was capable of performing. An engine built on the principle of those now used in England would not have kept on the rails a hundred yards, and yet this Bogie engine took us a distance of 58 miles, over such a road as I have described, in perfect safety.

At Montreal, Robert Stephenson was entertained (on August 19, 1853) by the citizens, with a banquet as magnificent as any that is mentioned in the records of that hospitable city. The speech made on the occasion by Robert Stephenson was the best he had ever made in public. He said :—

He had come to Canada upon a professional visit, and from what he had seen of the country, he was convinced that the present was but the commencement of a great railway system ; and he ventured to express his sincere desire, that Canada should avoid those errors in producing the system which other countries had committed with theirs. Canada was yet only on the threshold, and the proper laying out of the plan he thought of as much importance as the introduction of railroads themselves. He hoped that the legislature of this country would not do as those of some countries had done, for although there were difficulties in the way, yet here it appeared to him they were principally those of legislation. Canada, at present, had an unoccupied field before her, and much would depend upon the first step. If it were taken with judgment, other difficulties would be of comparatively little importance. If it were taken injudiciously, what seemed but a speck in the west, might become a thunder-cloud. The dangers might not be apparent to Canadians at present, but before he sat down he would endeavour to make them sensible of them. He had seen the rise and progress of railways in England, and those who had not could hardly appreciate the enormous losses occasioned by false legislation. Few, indeed, knew the rise and progress of legislation with respect to them. After obstacles raised in the first instance had been removed, the people became sensible of the immense advantages of the railway horse, and the question assumed a new phase. Then all places rushed into railway speculations. The country was tolerably well filled up with railway lines. Competition arose within the walls of parliament, not for interest, so much as for vanity. Parliamentary committees took into their consideration, not who was right and who was wrong, but entered upon questions entirely subsidiary, not at all connected with the profit of the lines, or the necessity for making them. The consequence was, that committees sometimes decided upon different lines, upon reasons entirely apart from their real merits, or the scientific questions involved in the details. There was one district through which it was proposed to run two lines, and there was no other difficulty between them than the simple rivalry, that, if one got a charter, the other might also. But here, where the committee might have given both, they gave neither. In another instance, two lines were projected through a barren country, and the committee gave the one which afforded the least accommodation to the public. In another, where a line was to be run, merely to shorten the time by a few minutes, leading through a mountainous country, the committee gave both; so that where the committee might have given both, they gave neither, and where they should have given neither, they gave both. Such a species of legislation was faulty, and he hoped it would not be imitated in Canada. There was, indeed, a committee, then sitting in England, the attention of which he had called to these facts. After lines were granted, the

competition which began within the walls of parliament continued when the lines came to be put in operation. People
said it was necessary to have competition for the benefit of the
public, that the whole country would be under the dominion of
a railway corporation, and competition was the only means of
checking it, and preserving moderate prices. Well, he could
say, upon the authority of the Board of Trade, and from his own
knowledge, that, since competing lines commenced, out of 300
millions of pounds expended, 60 millions had been wasted; that
is, in duplicate lines. But, in order to mark the inconsistency
of the proceedings in railway legislation, when the London and
Birmingham was asked for the feasibility of the route was doubted;
great difficulties were suggested as being in the way; engineers
were called in to decide everything in opposition to it; the
estimates were disputed and doubted; it was maintained that
the company ought to prove the traffic that was to go over
it, and that 6 or 8 per cent. were to be obtained on the money
invested; in fact, a most paternal part was taken in the project.
Before parliament granted the charter, before the people were
allowed to expend their own money, they were asked to prove
the traffic, and the profit, and show a regular contract that
the work was to be done within the estimate. The people
clamoured for competition, and parliament granted the expenditure of two capitals. At that time it was believed that
competition would compel them to carry passengers almost
for nothing. But, what was the result? The opulent were
struck down, and the poor were reduced to penury. Nothing
but the resources of British commerce could have sustained
such a shock. These serious difficulties of legislation it might
be difficult to escape from in England, but they should be
weighed well before they were brought into Canada, either
by the present or any future government. It was said, that
all was right, that the public gained what the proprietors lost.
But the public had *not* gained. Capital was absorbed and
diverted from other profitable employment. Lines had been
located, which never would have been built, had a directing
genius presided over the chartering of them; and he did not
envy the man who could glory in one part of the community
prospering by the ruins of another. This error had been most

disastrous in England, and he hoped it would not be committed in Canada. Competition had answered no useful purpose. Like poverty, a mother of invention, competition had invented the remedy—amalgamation. There is an instance, where, including water communications, there are five competing lines. The result from competition was, that directors laid their heads together and raised tolls to the highest; but in Canada, where there was no competition nothing was to be done but to develop resources, and make the highest possible profits. In Belgium, which had employed one or two men to lay out the country, so as to obtain the greatest amount of accommodation with the least expenditure of money, the speed obtained was not perhaps so great as that in England; but it would bear comparison with that in any other country. The minor points of the country were filled up according to the original design, and all conducted with economy. In France no lines were allowed to be made unless they were called for, and made part of a great system. France was slow, but in other respects bore favourable comparison with Great Britain, where competition had marred the whole scheme. Switzerland was a collection of cantons, each preferring its own interests as to gauge and building without reference to the whole. They decided to send to England for an engineer to design for them a net-work of railways, and he (Mr. Stephenson) had the honour to be called in, and assisted in designing their railways from end to end, and capital was now flowing in, the country being satisfied because there were no rival lines; and there was no doubt of their completion. As Canadians wanted English capital, he advised them to be very guarded in the system they should adopt, especially in regard to reckless competition. He would for himself hesitate to recommend his friends in England to invest their money in Canadian railways, if reckless competition were allowed, for nothing but loss and confusion could result from such a course. Belgium, France, and Switzerland, all possessed a great advantage over England in having no rival lines, and in having laid out their main lines for the benefit of the whole country, rather than allowing to any town, or any portion of the country, a preponderating voice in their location. They had established from end to end of their countries a system in harmony with itself. They could not prevent portions of country having their railways, but such railways were too small to interfere with the great design. It was the interest of the whole country that was involved, not that of individuals, nor. of particular localities. Canadians ought to have a system which would work in all harmony. What was to be gained by ruinous competition? If passions and animosities were brought into play, nothing could result but destruction of property and loss of life too. He could not offer a better comparison between a system judiciously designed and a wild competition, than the steamboat he was on board on the previous night. He found that vessel was propelled by two engines, each separate from the other, and he would suppose that they were under the command of two rival companies. These would have their opponents and friends on board ; and he would suppose that the engineers quarrelled : what progress would they have made ? Would they have reached Three Rivers ? He had one word more to say. He could not leave the city without saying something on the all-important subject of the bridge. The St. Lawrence was a most magnificent river, *and he had received abundant information respecting it from Mr. Alexander Ross, who had been in the country several months ago, as to what the bridge ought to be. It was a startling project and at the time he had no idea of having anything to do with it.* But having studied the admirable report of Mr. T. C. Keefer, and the philosophical paper drawn up by Mr. Logan, he could understand from what had been written the general circumstances to be considered. If possible, he would be in Canada again at the breaking up of the ice. Still they might be anxious to learn his impressions about the bridge, and he would proceed to state them. According to his ideas there was no difficulty to be apprehended from the ice. From calculations he had made since his arrival in Montreal, he was satisfied that no shove that could come against the piers would be sufficient to displace them. The people here had no doubt seen the formation of the ice in winter, and its breaking up in the spring. He had himself noticed these stupendous and beautiful phenomena, of course, but he had not seen them as they are at Montreal. He would say, however, that there were facts attending the breaking up of ice, which were susceptible of mathematical demonstration, and the pressure of the shove was also capable of being valued.   It was therefore as easy as the addition of two and two to estimate the pressure necessary to retain the piers in their places.   He had been told that the ice was piled thirty and forty feet high, but its pressure could be easily withstood.   He had calculated that if ice were piled twice the height it would not disturb a well-constructed bridge.   So far, therefore, he would say for its permanency.   He would also perhaps be expected to say something about its site.   The minds of the citizens were agitated on the question whether it should be below or above the town.   He thought it his duty not to conceal his opinion on this point, nor to permit the agitation to go on, as if he hesitated on the point.   It was a difficult and important subject, but he had brought his best judgment to bear upon it, and having no interest in it one way or another, he was in a situation to make an independent selection.   Whether the place was to be that laid out by Mr. Keefer or not, he was satisfied it must be above the island, and not far below Nun's Island.

On his return to England Robert Stephenson sustained a great and irreparable loss in the death of his brother-in-law Mr. John Sanderson.   The brother of Mrs. Stephenson (Robert's ' Fanny ' of the old Newcastle days) had for years been one of his closest personal friends, dwelling in his house, and exercising a salutary supervision over his pecuniary arrangements, both in Gloucester Square and Great George Street.   Robert had so strong an affection for him, that after his death, he could not for months endure the solitude of Gloucester Square.   Closing the house, he took apartments in ' Thomas's Hotel,' Berkeley Square, and did not return to his residence till twelve months of mourning had expired.

While Robert Stephenson undertook the chief command in the construction of the St. Lawrence Bridge, Mr.

Alexander Ross was united with him as co-engineer, to carry out on the spot all that very important part of the work which the engineer-in-chief could not personally superintend. The positions and mutual relations of the two engineers were accurately defined in the deed of contract made September 29, 1853, between the Grand Trunk Railway Company of Canada, of the one part, and William Jackson, of Birkenhead, and Samuel Morton Peto, Thomas Brassey, and Edward Ladd Betts, all of London, in England, contractors, of the other part. 'The contractors,' runs the language of the deed, ' will make, build, and construct the said tubular bridge over the said River St. Lawrence, at or near Montreal, according to the plans, sections, and specifications prepared and drawn by Robert Stephenson, of London, aforesaid, Civil Engineer, M.P., and Alexander M'Kenzie Ross, of Montreal, C.E.' The deed then requires : ' The Bridge when completed to be in perfect repair, and of the best and most substantial character, and to be approved of by the said Robert Stephenson.' It further provides, ' that in case of the death, refusal, or inability to act of the said Alexander M'Kenzie Ross, another engineer shall from time to time be appointed by the said Robert Stephenson, in place of the said Alexander M'Kenzie Ross, and who shall have all the powers of the said Alexander M'Kenzie Ross. And in the event of the death, or refusal, or inability of the said Robert Stephenson, then all things then remaining to be done by the said Robert Stephenson shall be done by an eminent civil engineer, to be appointed by the president for the time being of the Institution of Civil Engineers in England, upon the requisition of the parties hereto, or either of them.' Thus, after completion of the work,

Robert Stephenson's approval—given on his sole and undivided responsibility—was required before the contractors should be held to have discharged their engagements. In case of the death, refusal, or retirement of the resident engineer, it devolved on the engineer-in-chief to appoint his successor; whereas, in case of Robert Stephenson's death, refusal, or retirement, Mr. Ross was not qualified to nominate an engineer to occupy the vacated place, but was required to act with the nominee of the President for the time being of the Institution of Civil Engineers of England.

Marking yet more strongly the high control to be exercised by the engineer-in-chief, a subsequent passage of the contract runs, ' that if any question or difference of opinion shall arise between the parties hereto, as to this agreement, or any matter connected therewith or arising thereout in any way, &c., it shall be referred to the absolute decision of the said Robert Stephenson, as sole arbitrator; and the decision of the said Robert Stephenson shall be binding and conclusive upon both parties as to the question or difference of opinion so referred to him.'

From the time of the construction of the London and Birmingham Line, when he acted so frequently as arbitrator between the company and their numerous contractors, Robert Stephenson was constantly referred to in the disputes of business men. He was no good friend to lawyers. The amount of litigation he prevented by amiable counsel would almost justify his memory being held in abomination in Chancery Lane.

In accordance with his agreement with the Grand Trunk Railway Company of Canada, he had all the details of the St. Lawrence Bridge settled under his supervision, most of them being worked out in his office in Great George Street, under the immediate superintendence of his cousin, Mr. George Robert Stephenson. The whole of the iron work was designed in Great George Street, was constructed in England, and was shipped to Canada with detailed instructions, so that the engineers on the other side the Atlantic had little else to do with that part of the work except to put the pieces together in the manner directed.

Robert Stephenson did not make another visit to Canada, as nothing occurred in the execution of his designs to demand his presence.

CHAPTER VIII.

THE GREAT VICTORIA BRIDGE OVER THE RIVER ST. LAWRENCE IN CANADA.*

One of Mr. Stephenson's Last Works—Line of Lakes and River St. Lawrence—Difficulties of the Navigation—Introduction of Railways into Canada—The Grand Trunk Railway—Engineering Problem in the Design of the Bridge—Phenomena of the Ice—Early Proposals for a Bridge—Mr. A. W. Ross—Mr. Stephenson consulted—Joint Report —Mr. Stephenson visits Canada and reports again to the Directors— Surveys—Letting of the Contract—Iron-work—Mr. G. R. Stephenson —Controversy on the fitness of the Design—Mr. Stephenson's Views —Description of the Bridge—Site—Piers—Tubes—Erection of the Bridge—Foundations—Caissons—Shortness of the Working Season— Contrivances to save Time—Inspection of the Bridge—Opening by the Prince of Wales—Difficulties overcome.

THIS was one of the last works on which Mr. Stephenson was engaged; it was opened a few months after his death. It is the largest bridge in the world, and its design has involved engineering problems of a peculiar and unusual nature.†

The Victoria Bridge carries the Grand Trunk Railway of Canada across the River St. Lawrence, at a point near Montreal, where it is nearly two miles wide.

---

* This chapter is contributed by Professor Pole.

† An elaborate work, profusely illustrated, descriptive of the Bridge, has been published by Mr. James Hodges, the engineer to the contractors. Weale, London, 1860. There is also a cheap and useful little 'Handbook of the Victoria Bridge,' written by Mr. F. N. Boxer, Architect and Civil Engineer, and published by Hunter, Montreal, 1860.

Free use has been made of both these works in some parts of the present notice.

One of the most prominent features in the map of North America is the long line of water communication, which, commencing with the magnificent series of Lakes Superior, Michigan, Huron, Erie, and Ontario, descends from the latter of these by the River, the Estuary, and the Gulf of St. Lawrence, to the Atlantic Ocean.

From Lake Superior to a little below Lake Ontario this water line forms the frontier between Canada and the United States of America, but the boundary then retreats inland, on the south side, so that the lower part of the river, namely, from the 45th degree of latitude downwards, runs entirely within the British territory, and thus separates two parts of our colonial possessions which are of great richness and commercial value. On this part of the river also are situated the two large and important towns of Quebec and Montreal, the former at about 250 miles, and the latter at about 400 miles, above the mouth of the river, or rather above the Point des Monts, where the estuary may be considered to end and the gulf to begin.

To this grand river, and the fine extent of inland navigation above it, Canada is no doubt largely indebted for her prosperity and growth. But for nearly half the year the St. Lawrence is sealed up by frost; and during this time not only is all water traffic suspended between the different parts of the country, but Quebec and the other ports of the river and the lakes are deprived, so far as water communication is concerned, of that easy communication with the ocean on which their commercial prosperity so much depends.

But there are also difficulties attending the upper part of the navigation, even in summer. The Welland Canal,

cut to form a navigable communication past the Niagara
Falls, is of such contracted dimensions, that although
vessels of 700 or 800 tons burthen can with ease get
up to Lake Ontario, none greater than about 300 tons
can reach the upper chain of lakes; and therefore, at
this point, trans-shipment becomes necessary, both for
the import and the export trade.

The obvious remedy for these disadvantages was a
system of railways; and accordingly at an early period
the inhabitants expressed a wish to avail themselves of
this mode of providing for their traffic. The first effort
that took a practical shape was a line of railway intended to connect the River St. Lawrence, at a point
opposite Montreal, with the great Atlantic harbour of
Portland, distant about 230 miles to the SE. This
line was called the 'Atlantic and St. Lawrence Railway;'
it had to cross the frontier, and thus was partly Canadian and partly belonging to the United States. It was
commenced about 1849; and by the middle of 1852,
at which time its proprietorship changed hands, two
lengths of it were opened, namely, from the St. Lawrence
eastward to Sherbrooke, a distance of 80 miles; and
from Portland northward about 90 miles, leaving a
break in the middle of about 60 miles unfinished.

By this time, however, two other railways had been
projected, one called the Quebec and Richmond Railway, commencing at Richmond on the last-named line,
and passing eastward to Quebec, a distance of about
85 miles; the other on the north bank of the River
St. Lawrence, extending from Montreal westward to
Kingston, about 170 miles, and called the Montreal and
Kingston Railway. At the time referred to, the former

of these was in progress, and for the latter, powers to construct had been obtained.

But it soon became evident that a more comprehensive scheme of railways was desirable, and the subject having excited interest in England during the summer of 1852, an examination of the country was made, at the request of the Provincial Government of Canada, by a firm of English contractors, with the view of carrying out such a project. The information they obtained led them to promote the formation of a company, with the object, first, of purchasing and completing the lines partly made or sanctioned; secondly, of adding other important extensions; and thirdly, of uniting the whole into one great system, by a gigantic bridge across the great water chain.

This vast enterprise is now completed, under the name of the Grand Trunk Railway of Canada; and it comprises upwards of 1200 miles of line. About one-half of this is on the north side of the River St. Lawrence, extending from Montreal westward as far as Lake Huron; the remainder is on the south side, and has two main branches, one, the original Atlantic and St. Lawrence line, passing southward to Portland; the other, the original Quebec and Richmond line, extending eastward to Quebec. There is also a further extension, which, passing down the south bank of the St. Lawrence, ends at present at a place called Rivière du Loup. This latter branch was made partly to open out the south-eastern provinces of Canada, and to develope the great agricultural resources of that rich district; but it had also reference to a far more important ultimate purpose.

One of the main objects proposed to be attained by

the introduction of railways into the colony was to open
a communication between the interior of the country
and the sea, during that long period of the year when
the St. Lawrence is closed by frost. At present, the
Grand Trunk Railway only accomplishes this by its
connexion with the United States harbour of Portland;
but it is evident that circumstances might arise which
would render this mode of communication unavailable;
and therefore it would be highly expedient to complete
the new roadway to the sea entirely through British
territory; a measure, the expediency of which has often
been forcibly urged. The most obvious means of attain-
ing this end would be to continue the line last mentioned
from the present temporary eastern terminus, about 400
miles further in a south-easterly direction, namely, across
New Brunswick and Nova Scotia to the fine British port
of Halifax, which would put the whole of Canada in
communication with the ocean, at all times of the year,
in a manner independent of any foreign power.

The Grand Trunk Railway is, however, even in its
present incomplete state, a remarkable national work.
It traverses British North America in one unbroken line
for half the breadth of a continent; it connects and
associates together the several wide-lying dependencies of
the British Crown; it affords throughout the year to the
inhabitants of the wonderful valley through which it
passes constant means of communication and transport
which nature had denied them; it opens out the rich
prairie country of the Far West; and it unites the whole
of these vast regions with the neighbouring territories of
America, and with the great maritime pathway of the
world.

Our attention must now, however, be confined to the great bridge, by which the connection between the two main divisions of the line has been secured. The necessity for such a bridge became urgent when the railways had made some progress, and Mr. Stephenson's own remarks on this point will be hereafter inserted. Suffice it here to say that from the head of Lake Superior to the Atlantic Ocean, a distance of more than 1500 miles, there was not any bridge across the great water chain, excepting at the Niagara gorge; and that, therefore, the key to the province would be obviously in other hands, if the railway communication could not be completed between the North and the South sides.

The engineering problems to be solved in the erection of such a bridge near Montreal, were of a peculiar kind. Although, at the point where it was desired to cross, the river was very wide, and the current was strong, yet the water was comparatively shallow, and it seemed unlikely that the nature of the bed would offer much trouble with the foundations. And since, after the experience gained in the Britannia and Conway Bridges, the use of iron for spanning large openings had become easy, the construction of a railway bridge would have involved nothing beyond a question of expense, had it not been for a difficulty peculiar to the locality, and one scarcely known in English engineering—the extraordinary effects produced upon the river by the great severity of the Canadian winter.

To explain these phenomena it will be sufficient to quote the description of them given by Mr. W. E.

Logan, F.G.S., in the Proceedings of the Geological Society of London for 1842. He says :—

The frosts commence about the end of November, and a margin of ice of some strength soon forms along the shores; and wherever the water is still, it is immediately cased over. The first barrier completed across the river, below Montreal, is usually formed about Christmas, at the entrance of Lake St. Peter, where the St. Lawrence is divided into a multitude of channels by low alluvial islands. This barrier is rapidly increased by extensive fields of drift ice, enormous quantities of which are heaped upon or forced under, the stationary mass.

The space left for the water to flow being thus greatly diminished, a perceptible rise in the river takes place, and by the time that the ice becomes stationary at the foot of St. Mary's current, opposite Montreal, the waters in the harbour have usually risen several feet, and as the packing rapidly proceeds, they soon attain the height of 20 and sometimes 26 feet above the summer level.

It is at this period that the grandest glacial phenomena are presented. In consequence of the *packing* and *piling* of the ice, as well as the accumulation of the moistened snow of the season, and the freezing of the whole into a solid body, sometimes more than 20 feet thick, the water suddenly rises, and lifting a wide expanse of the entire covering of the St. Lawrence, urges it forward with terrific violence, piling up the rended masses on the banks of the narrower parts of the river, to the height of 40 or 50 feet.

In front of Montreal is a newly-built revêtement, the top of which is 23 feet above the summer level of the river; but the ice broken by it accumulates on the surrounding terrace, and before the wall was erected the adjacent buildings were endangered, the ice sometimes breaking in at the windows of the second floor, even 200 feet from the margin of the river.

In one instance, a warehouse of considerable strength and magnitude having been erected without this protection, the great moving sheet of river ice pushed it over as if it had been a house of cards; and in another case, where a similarly situated and equally extensive warehouse, 4 or 5 stories high, had been provided with a range of oaken piles, placed at an angle of less than 45°, the drift ice rose up the inclined plane, and after meeting the walls of the building, fell back, and formed, in a few minutes, an enormous but protecting rampart. In some years the ice accumulated nearly as high as the roof of the warehouse.

Several of these grand glacial movements [locally termed 'shovings'] take place, sometimes at intervals of many days, but occasionally of only a few hours, the permanent setting being indicated by a longitudinal opening of considerable extent in some part of St. Mary's current. This opening, which is never afterwards frozen over, even when the temperature is 30° below zero of Fahrenheit, is due to the water having formed a free subglacial as well as superficial passage, in consequence of its own action and the cessation in the supply of drifting ice.

From this period the waters gradually subside, but seldom or never to their summer level; and when they have attained their minimum, the trough of the St. Lawrence exhibits a glacial landscape of undulating hills and valleys of ice.

This description will make it clear that the great engineering difficulty to be overcome in building a bridge across the river was, so to establish the piers (of which a large number were required), that while they should offer the least possible resistance to the progress of the ice down the stream, they should be strong enough safely to withstand the enormous and almost unprecedented destructive force to which they must be exposed from its violent action.

No doubt the idea that it would be advantageous to bridge over the St. Lawrence must have occurred to the Canadians at a very early period of the history of the colony; but we find no published intimation that such a scheme was considered practicable until, in June 1846, discussions were raised as to the proper site for the

terminus of the Atlantic and St. Lawrence Railway, then
in progress. At this time, the Honourable John Young,
an energetic citizen of Montreal, and one of the railway
directors, published a newspaper article, pointing out
that the passage of the river by the railway would afford
great facilities for the traffic, and expressing a confident
opinion that the erection of a bridge would be ' perfectly
practicable,' at a certain point named, coinciding very
nearly with the site on which the bridge now stands.

The suggestion seems to have made a favourable
impression, for in September of the same year the
directors of the Atlantic and St. Lawrence Railway
authorised Mr. A. C. Morton, their engineer, to cause a
survey to be made for the purpose of ascertaining the
practicability of building a bridge and of estimating its
probable cost. Mr. Morton took soundings, but the
nature of his report does not appear to have been
made public. In the meantime, a committee of
citizens of Montreal was formed to investigate the same
subject, and they employed Mr. E. H. Gay, of Pennsylvania, then engineer of the Columbia and Philadelphia
Railway, to make another survey. His report was
given in December 1846, accompanied by plans and
estimates. He disapproved of the line selected by Mr.
Morton, but considered it practicable to build a bridge
across another site, at the moderate expense of about
525,000 dollars ; giving, however, no provision for the
navigation of the river by masted vessels.

About this time a general commercial depression seems
to have prevailed throughout the province, and nothing
further was done till June 1851, when, on the promotion of the railway from Montreal to Kingston, which

was proposed to join the Atlantic and St. Lawrence line, the two companies agreed that another survey should be made for a bridge; and accordingly, Mr. Thomas C. Keefer, the engineer of the Kingston line, was commissioned to undertake it. He proceeded with the work in this and the following year, but his progress was delayed for want of funds. He wrote a report on the subject, which was published in 1853,* but the date of its presentation is not recorded; it was an able document, containing much valuable information, and Mr. Stephenson expressed a high estimation of Mr. Keefer's labours.

Meantime, however, the interest in the work had passed into other hands. The St. Lawrence and Atlantic line had become a part of the more comprehensive scheme, the Grand Trunk Railway; and in September 1852 the Kingston Company also waived their charter in favour of the same great enterprise, on condition that the new proprietors would construct the bridge.

When the examinations of the country were made, in the spring of 1852, with a view to the formation of the Grand Trunk Railway Company, the promoters had engaged as their engineer Mr. A. M. Ross, who had previously been one of the resident engineers under Mr. Stephenson, on the Chester and Holyhead Railway. He had charge of the masonry of the Conway Tubular Bridge, and Mr. Stephenson had a high opinion of his skill in that department of engineering. The question of the

---

* Report on a survey for the Railway Bridge over the St. Lawrence at Montreal, surveyed in 1851-52, by order of the committee of the Montreal and Kingston Railway. Hon. John Young, chairman; Thos. C. Keefer, engineer. Montreal, John Lovell, 1853.

bridge over the St. Lawrence formed naturally one of the most prominent subjects of enquiry, and in July 1852, Mr. Young, who has been already mentioned as the original projector of the bridge, and who had ever been its most energetic promoter, took Mr. Ross to examine the various points of crossing that had been proposed; and it is stated that, when near the site subsequently adopted, Mr. Ross suggested that the iron tubular beam principle would be applicable for the superstructure. In October 1852, immediately after the Grand Trunk Company had taken the matter into their own hands, surveys were commenced on their behalf. It was probably at this time that Mr. Keefer handed over, for the use of the new company, the extensive and valuable information he had obtained. About the end of this year Mr. Ross returned to England.

The great importance of the bridge, the large expenditure it involved, the various opinions that existed as to its practicability, and the great difficulties and risks connected with its construction in such a position, decided the board of directors, previous to bringing it before the public, to consult Mr. Robert Stephenson, whose high opinion it was deemed of great importance to obtain.

With this view Mr. Ross, after his return to England, laid before Mr. Stephenson all the information collected both by Mr. Keefer and by himself. The subject had some months' careful consideration, and after many tentative plans and estimates had been prepared, on March 18, 1853, Mr. Stephenson communicated to the Hon. John Ross, Speaker of the Canadian House of Assembly, a design similar in its principal features to that ultimately carried out.

Mr. A. M. Ross returned to Canada in April, and soon after his arrival, the same design was submitted to the Board of Railway Commissioners at Quebec, with a joint report explaining its general features, and of which the following is a copy : *—

Grand Trunk Railway, Champ de Mars, Montreal:
June 6, 1853.

To the Honourable the Board of Railway Commissioners,
Quebec.

GENTLEMEN,—We beg herewith to transmit for your approval, a design for the proposed Victoria Bridge for carrying the railways across the River St. Lawrence, at this place.

On the map accompanying this report the precise locality is clearly defined : the figures marked upon the map indicate the depth of summer water, determined by soundings carefully taken; the shoals, which are numerous and intricate, are also outlined as nearly as they could be ascertained.

From an inspection of the map it will be seen that the site selected for the bridge embraces as wide a range of deep water as can be obtained by any line crossing the river, in the vicinity chosen as the most eligible for this important structure, and where the width across from bank to bank is 8,600 feet, the deep water channel occupying about one-seventh of this width.

The abutments of the proposed structure are placed 6,588 feet apart, and the piers (twenty-four in number) occupy 450 feet of this space, leaving 6,138 feet clear waterway, which is equal to 93 per cent. of the whole, having an average summer depth of 9 feet water, the navigable channel being $15\frac{1}{2}$ feet deep.

It is proposed to fill up the intervening space between the abutments and the shores on either side (700 feet in length on the St. Lambert, and 1,300 feet on the Point St. Charles side) with solid embankment composed of stone. The form of this embankment is delineated upon the drawing.

---

* This report is dated at Montreal; but as Mr. Stephenson was at that time in England, it is probable that he authorised Mr. Ross, who had then returned to Canada, to affix his signature.

The piers are proposed to be built of solid masonry, of such form and proportions as will be in every way calculated to withstand any pressure to which they may be liable from the moving ice.

An explanatory diagram of these is shewn upon the drawing.

The superstructure is proposed to be of wrought iron, constructed in every respect on the same principle as the Britannia Bridge over the Menai Straits, on the Chester and Holyhead Railway, and in uniform spans of 242 feet, excepting that over the navigable channel, which is intended to be 330 feet.

The strength is calculated to resist four times the actual load to be sustained, and equal to ten times the moving load, reckoned at one ton to the lineal foot.

The clear headway above summer water is placed at 60 feet for the whole width of the centre opening, a height which, from the best information we can obtain, is ample for the passage of any craft which can come down the rapids.

From the pier on either side of the centre opening, the height gradually diminishes at the rate of 1 in 130 to the extreme end of the tubes, and from this point falls towards the shores at the rate of 1 in 100, to suit the local requirements connected with the railways on either side of the river.

The leading characteristics of the design may be stated briefly as follow : That only one-fifteenth of the space between the abutments is occupied by the supporting piers. That the piers are of the form best suited for withstanding any force to which they may be exposed from the moving ice, and for severing the floating masses in their progress. That a wide deepwater channel is selected as affording the greatest security in reference to the passage of the ice, and that the materials of construction are of that permanently enduring character as will require a minimum amount for efficient maintenance.

With a due regard to every consideration involved in this important measure the accompanying plans are respectfully submitted for approval.

(Signed)   R. STEPHENSON.
A. M. ROSS.

The Railway Board, on receiving this report, instructed Mr. Killaby, the Assistant Commissioner of Public Works, to examine the plans, of which he reported his entire approval. He discussed at some length the height of headway proposed, as some persons were of opinion it should not be less than 100 feet; but he shewed it was so improbable that vessels requiring this height of headway could descend by the rapids from the lakes above, that it would not be requisite to inflict on the bridge the great and permanent injury of raising it so high. Mr. Killaby's report was adopted by the Government, and the proposed bridge formally approved; and an intimation of the approval was sent to Mr. Stephenson and Mr. Ross on August 19, 1853.

Although, however, the design of the bridge was thus arranged, with tolerable precision, it was felt that the weight of Mr. Stephenson's authority would be much enhanced if he actually visited the site, and took personal cognizance of all the various circumstances affecting the measure. He accordingly left England about the middle of July, and remained in Canada till September. This visit enabled him to enter much more fully into the details of the subject than before, and his more matured views upon it were expressed in a letter to the directors, dated May 2, 1854, which is of sufficient importance to warrant its insertion entire :—

24 Great George Street, Westminster :
May 2, 1854.

GENTLEMEN,—Absence from England, and other unexpected circumstances, have prevented my sooner laying before you the results of my visit to Canada last autumn, for the purpose of conferring with your engineer-in-chief, Mr. Alexander Ross,[*]

---

[*] Mr. Ross was engineer-in-chief to the railway, with the general works of which Mr. Stephenson had nothing to do.

respecting the Victoria Bridge across the River St. Lawrence, in the vicinity of Montreal.

The subject will naturally render itself into three parts, viz:—

First—The description of bridge best adapted for the situation.

Second—The selection of a proper site.

Thirdly—The necessity for such a structure.

Regarding the first point, I do not feel called upon to enter on a discussion of the different opinions which have been expressed by engineers, both in England and America, as to the comparative merits of different classes of bridges, and more especially as between the suspension and tubular principles, when large spans become a matter of necessity. It is known to me that in one case in the United States a common suspension bridge has been applied to railway purposes; but from the information in my possession, from a high engineering authority in that country, the work alluded to can scarcely be looked upon as a permanent, substantial, and safe structure. Its flexibility, I was informed, was truly alarming, and although another structure of this kind is in process of construction near Niagara, in which great skill has been shown in designing means for neutralising this tendency to flexibility, I am of opinion that no system of trussing applicable to a platform suspended from chains will prove either durable or efficient, unless it be carried to such an extent as to approach in dimensions a tube, fit itself for the passage of railway trains through it. Such bridge may doubtless be successfully, and perhaps with propriety, adopted in some situations; but I am convinced that even in such situations, while they will in first cost fall little short of wrought-iron tubes, they will be more expensive to maintain, and far inferior in efficiency and safety.

I cannot hesitate, therefore, to recommend the adoption of a tubular bridge, similar in all essential particulars to that of the Britannia over the Menai Straits in this country; and it must be observed that, the essential features being the same, although the length much exceeds that of the work alluded to, none of the formidable difficulties which surrounded its erection will be involved in the present instance. In the Britannia, the two larger openings were each 460 feet, whereas in the proposed

Victoria there is only one large opening of 330 feet, all the rest being 240 feet. In the construction of the latter, there is also every facility for the erection of scaffolding which will admit of the tubes being constructed in their permanent position, thus avoiding both the precarious and expensive process of floating, and afterwards lifting the tubes to the final level by hydraulic pressure.

In speaking of these facilities, it is a most agreeable and satisfactory duty to put on record that the Government Engineering Department has, throughout the consideration of this important question, exhibited the most friendly spirit, and done everything in its power to remove several onerous conditions, which were at one time spoken of as necessary, before official sanction would be given for the construction of the work.

On my arrival in Canada, I found that Mr. A. M. Ross had collected so much information bearing on the subject of the *site* of the bridge, that my task was comparatively an easy one.

Amongst the inhabitants of Montreal, I found two opinions existing on this point—somewhat conflicting: the one side maintaining that the river should be crossed immediately on the lower side of the city, where the principal channel is much narrower than elsewhere, and where also the island of St. Helen's would shorten the length of the bridge; the other seeming to be in favour of crossing a little below Nunn's Island.

Sections of the bed of the river at both points had been prepared, and a careful study of these left no doubt on my mind that the latter was decidedly the one to be adopted.

In addition, however, to the simple question of the best site for the construction of a bridge across the St. Lawrence, my attention was specially called to the feasibility of erecting and maintaining such a structure during the breaking up of the ice in spring, when results take place which appear to every observer indicative of forces almost irresistible; and, therefore, such as would be likely to destroy any piers built for the support of a bridge. I have not myself had the advantage of witnessing these remarkable phenomena, but have endeavoured to realise them in my mind as far as practicable, by conversation with those to whom they are familiar; and, in addition to this, I have read and studied with great pleasure an admirable and most graphic description by Mr. Logan, of the whole of the varied conditions of the river, from the commencement of the formation of ice to its breaking up and clearing away in spring. To this memoir I am much indebted for a clear comprehension of the formidable tumult that takes place at different times amongst the huge masses of ice on the surface of the river, and which must strike the eye as if irresistible forces were in operation, or such as, at all events, would put all calculations at defiance.

This is no doubt the first impression on the mind of the observer; but more mature reflection on the subject soon points out the source from which all the forces displayed must originate.

The origin of these powers is simply the gravity of the mass occupying the surface of the water, with a given declivity up to a point where the river is again clear of ice; which, in this case, is at the Lachine Falls. This is unquestionably the maximum amount of force that can come into play; but its effect is evidently greatly reduced—partly by the ice attaching itself to the shores, and partly by its grounding upon the bed of the river. Such modifications of the forces are clearly beyond the reach of calculation, as no correct data can be obtained for their estimation; but if we proceed by omitting all consideration of those circumstances which tend to reduce the greatest force that can be exerted, a sufficiently safe result is arrived at.

In thus treating the subject of the forces that may be occasionally applied to the piers of the proposed bridge, I am fully alive to the many other circumstances which may occasionally combine in such a manner as apparently to produce severe and extraordinary pressure at points on the mass of ice or upon the shore; and, consequently, upon the individual piers of a bridge. Many inquiries were made respecting this particular view, but no facts were elicited indicative of forces existing at all approaching to that which I have regarded as the source and the maximum of the pressure that can at any time come into operation affecting the bridge.

I do not think it necessary to go into detail respecting the precise form and construction of the piers, and shall merely state that in forming the design care has been taken to bear in

mind the expedients which have hitherto been used and found
successful in protecting bridges exposed to the severe tests of a
Canadian winter, and the breaking up of the ice of frozen
rivers.

I now come to the last point, viz. the *necessity* for this large
and costly bridge.

Before entering on the expenditure of £1,400,000 upon
one work in any system of railways, it is of course necessary to
consider the bearing which it has upon the entire undertaking
if carried out, and also the effect which its postponement is likely
to produce.

These questions appear to me to be very simple, and free from
any difficulty.

An extensive series of railways in Canada, on the north side
of the St. Lawrence, is developing itself rapidly ; part of it is
already in operation, a large portion fast progressing, and other
lines in contemplation, the commencement of which must
speedily take place.

The commerce of this extensive and productive country has
scarcely any outlet at present but through the St. Lawrence,
which is sealed up during six months of the year, and therefore
very imperfectly answers the purposes of a great commercial
thoroughfare.

Experience, both in this and other countries where railways
have come into rivalry with the best navigable rivers, has
demonstrated, beyond the possibility of question, that this new
description of locomotion is capable of superseding water
carriage wherever economy and despatch are required ; and
even where the latter is of little importance, the capabilities of
a railway, properly managed, may still be made available simply
for economy.

The great object, however, of the Canadian system of railways
is not to compete with the River St. Lawrence, which will continue to accommodate a certain portion of the traffic of the country,
but to bring those rich provinces into direct and easy connection
with all the ports on the east coast of the Atlantic, from Halifax
to Boston, and even New York—and consequently through these
ports nearer to Europe.

If the line of railway communication be permitted to remain

severed by the St. Lawrence, it is obvious that the benefits which
the system is calculated to confer upon Canada must remain in
a great extent nugatory and of a local character.

The province will be comparatively insulated, and cut off from
that coast to which her commerce naturally tends; the traffic
from the west must either continue to adopt the water communication, or, what is more probable — nay, I should say,
certain—it would cross into the United States by those lines
nearly completed to Buffalo, crossing the river near Niagara.

No one who has visited the country and made himself acquainted only partially with the tendencies of the trade which
is growing up on all sides in Upper Canada, can fail to perceive,
that if vigorous steps be not taken to render the railway communication with the eastern coast through Lower Canada uninterrupted, the whole of the produce of Upper Canada will find
its way to the coast through other channels; and the system of
lines now comprised in your undertaking will be deprived of
that traffic upon which you have very reasonably calculated.

In short, I cannot conceive anything so fatal to the satisfactory
development of your railway as the postponement of the bridge
across the river at Montreal. The line cannot, in my opinion,
fulfil its object of being the high road for Canadian produce
until this work is completed; and looking at the enormous
extent of rich and prosperous country which your system intersects, and at the amount of capital which has been already,
or is in progress or prospect of being expended, there is in my
mind no room for question as to the expediency—indeed, the
absolute necessity of the completion of this bridge, upon which,
I am persuaded, the successful issue of your great undertaking
mainly depends.

I am, gentlemen,

Yours faithfully,

(Signed)   ROBERT STEPHENSON.

To the Directors of the
Grand Trunk Railway of Canada.

Meantime, active preparations had been made for the
construction of the bridge. The detailed surveys had

been progressing all the year; the visit of Mr. Stephenson and his conferences with Mr. Ross on the spot had tended to settle more conclusively the details of the design ; and on September 29, 1853, the contract was let for the construction of the bridge.   The contractors were Messrs. William Jackson, Samuel Morton Peto, Thomas Brassey, and Edward Ladd Betts; and the contract sum for the entire work was £1,400,000.

The iron work of the superstructure had to be made in England, and the designs for this were entrusted by Mr. Stephenson to his cousin, Mr. George Robert Stephenson, who carried them out in all their details, and superintended the entire manufacture at the ' Canada ' iron works, Birkenhead.

Some material changes were made in the superstructure subsequently to the original design.   The first proposal was, in all the openings except three, to place the road on the *top* of the tubes, as Mr. Stephenson had done in the Egyptian bridges, particulars of which Mr. Ross took with him to Canada.   In the middle or navigable opening, in order to gain headway, the trains were to run *through* the tube, as in the Britannia Bridge, and the adjoining openings on each side were treated similarly, so as to form a continuous tubular girder of three spans. On further consideration, however, it was thought better to adopt this latter plan in all the tubes, and they were altered.

Other changes were also made in regard to the construction of the girders.   The central tube, being of larger span than the others, had been originally designed with a cellular top, like the Britannia and Conway Bridges; but Mr. Stephenson having gained more confidence in simpler means of giving rigidity, substituted plain plates, with stiffening bars of angle and T iron.

Further, in the original design, the tubes were connected together in lengths of four spans each; these were afterwards reduced to two spans each, at Mr. G. R. Stephenson's suggestion, in order to diminish · the tendency to roll down the incline.

About two years after the letting of the contract, and when the works were considerably advanced, a controversy arose as to the fitness of the design for the bridge. It was represented to the Directors that the plans adopted were extravagantly expensive, and that by using a different kind of girder for the superstructure, and different methods of founding the piers, a sufficient bridge might be erected for about one-fourth the cost.

Mr. Stephenson replied to these assertions at considerable length, in a report dated November 3, 1855, and the opinions of Mr. Brunel, Mr. Edwin Clark, and Mr. Ross, were also laid before the Directors in confirmation of Mr. Stephenson's views.

It appears that the Board were satisfied as to the propriety of the designs, and the works were allowed to go on. It is unnecessary, therefore, to enlarge further on these discussions, but the following passages from Mr. Stephenson's report throw so much light on his views as to warrant their insertion :—

It would evidently be unreasonable to expect that amongst professional men an absolute identity of opinion should exist, either in reference to the general design, or in many of the details of a work intended to meet such unusually formidable natural difficulties as are to be contended with in the construction of a bridge across the St. Lawrence.

You will remember that at the time I first entered upon the consideration of the subject, these difficulties were deemed by many well acquainted with the locality, and publicly stated by them, to be, if not insurmountable, at all events of so serious a character as to render the undertaking a very precarious one.

The information I received respecting these obstacles, when my attention was first drawn to this project, was so striking, that I reserved forming an opinion until I had visited the spot, had well considered all the detailed information which Mr. Alexander Ross had collected during several months' previous residence in the country, and had heard the opinion of many intelligent residents, regarding the forces exhibited by the movements of huge masses of ice during the opening of the river in spring.

The facts gathered from these sources fully convinced me that, although the undertaking was practicable, the forces brought into action by floating ice, as described, were of a formidable nature, and could only be effectively counteracted by a structure of a most solid and massive kind.

All the information which has been collected since I made my first report has only tended to confirm the impressions by which I was then guided.

For the sake of clearness and simplicity, the consideration of the design may be divided into four parts: first, the approaches; secondly, the foundations; thirdly, the upper masonry; and fourthly, the superstructure or roadway.

The approaches—extending in length to 700 feet on the south, or St. Lambert side, and 1,300 feet on the Point St. Charles side—consist of solid embankments, formed of large masses of stone, heaped up, and faced on the sloping sides with rubble masonry.

The up-stream side of these embankments is formed into a hollow shelving slope, the upper portion of which is a circular curve of 60 feet radius, and the lower portion, or foot of the slope, has a straight incline of 3 to 1; while the down-stream side, which is not exposed to the direct action of the floating ice, has a slope of 1 to 1. These embankments are being

constructed in a very solid and durable manner, and from their
extending along that portion of the river only where the depth
at summer level is not more than 2 feet 6 inches, the navigation is not interrupted, and a great protection is, by their means,
afforded to the city from the effect of the 'shoves' of ice which
are known to be so detrimental to its frontage.

Advantage has also been taken of the shallow depth of water
in constructing the abutments, which are each 242 feet in
length, and consist of masonry of the same description as that
of the piers, which I am about to describe; and from their being
erected in such a small depth of water, their foundations do
not require any extraordinary means for their construction.

The foundations, as you are aware, are fortunately on solid
rock, in no place at a great depth below the summer level of
the water in the river.

Various methods of constructing the foundations suggested
themselves, and were carefully considered; but without deciding
upon any particular method of proceeding, it was assumed that
the diving bell, or such modifications of it, on a larger scale, as
have been recently employed with great success in situations
not very dissimilar, would be most expedient. The contractors,
however, or rather the superintendent, Mr. Hodges, in conjunction with Mr. Ross, after much consideration on the spot,
devised another system of laying the foundations, which was by
means of floating 'coffer dams,' so contrived that the usual
difficulty in applying coffer-dams for rock foundations would
be, it was hoped, in a great measure obviated. When in
Montreal, I examined a model of this contrivance, and quite
approved of its application, without feeling certain that it
would materially reduce the expense of construction below that
of the system assumed to be adopted by Mr. Ross and myself,
in making the estimate. In approving of the method proposed
by Mr. Hodges, I was actuated by the feeling that the engineers
would not be justified in controlling the contractors in the
adoption of such means as they might consider most economical
to themselves, so long as the soundness and stability of the
work were in no way affected.

This new method has been hitherto acted upon with such
new modifications as experience has suggested from time to

time, during the progress of the work, and although successfully,
I learn from the contractors that experience has proved the bed
of the river to be far more irregular than was at first supposed
—presenting, instead of tolerably uniform ledges of rock, large
loose fragments, which are strewed about, and cause much
inconvenience and delay.

They are therefore necessitated to vary their mode of proceeding to meet these new circumstances ; and it may be stated
that all observations, up to this time, show the propriety, notwithstanding the difficulty with dams, of carrying the ashlar
masonry of the piers down to a solid rock.

We are now brought to the question as to the upper masonry.
This question is exceedingly important, since the cost of the
masonry constitutes upwards of 50 per cent. of the total estimated cost of the bridge and approaches. The amount of the
item of expenditure for the masonry is clearly dependent upon
the number of piers, which is again regulated by the spans
between them.

The width of the openings in bridges is frequently influenced,
and sometimes absolutely governed, by peculiarities of site.   In
the present case, however, the spans, with the exception of the
middle one, are decided by a comparison with the cost of the
piers; for it is evident that so soon as the increased expense in
the roadway, by enlarging the spans, balances the economy
produced by lessening the number of piers, any further increase
of span would be wasteful.

Calculations based upon this principle of reasoning, coupled
to some extent with considerations based upon the advantages
to be derived from having all the tubes as nearly alike as possible, have proved that the spans which have been adopted in
the present design for all the side openings, viz. 242 feet, have
produced the greatest economy.   The centre span has been
made 330 feet, not only for the purpose of giving every possible
facility for the navigation, but because that span is very nearly
the width of the centre and principal deep channel of the
stream.

It may perhaps appear to some, in examining the design,
that a saving might be effected in the masonry by abandon-
ing the inclined planes, which are added to the up side of

each pier, for the purpose of arresting the ice, and termed 'ice-breakers.'

In European rivers, and I believe in those of America also, these 'ice-breakers' are usually placed a little way in advance of, or rather above, the piers of the bridges, with a view of saving them from injury by the ice shelving up above the level of (frequently on to) the roadway.

In the case of the Victoria Bridge, the level of the roadway is far above that to which the ice ever reaches; and as the ordinary plan of 'ice-breakers,' composed of timber and stone, would be much larger in bulk, though of a rougher character, than those which are now added to the piers, I have reason to believe that they would be equally costly, besides requiring constant annual reparation. It was therefore decided to make them a part of the structure itself, as is now being done.

To convey some idea of the magnitude of ordinary 'ice-breakers' placed on the up-side of the pier, and to enable you to form some notion of their cost, I cannot do better than quote the following from the excellent report addressed to the Honourable John Young, by Mr. Thomas C. Keefer, whose experience in such matters, from long residence in the country, entitles his opinion as to the proper character of such works to confidence:—

'The plan I have proposed contemplates the planting of very large "cribs" or wooden "shoes," covering an area of about one-fourth of an acre each, and leaving a clear passage between them of about 240 feet—a width which will allow ordinary rafts to float broadside between them. These "islands" of timber and stone will have a rectangular well left open in the middle of their width, toward their lower ends, out of which will rise the solid masonry towers, supporting the weight of the superstructure, and resting on the rocky bed of the river. This enclosure of solid crib-work all round the masonry, yet detached from it, will receive the shock, pressure and grinding of the ice, and yield to a certain extent, by its elasticity, without communicating the shock to the masonry piers. These cribs, if damaged, can be repaired with facility, and, from their cohesive powers, will resist the action of the ice better than ordinary masonry. During construction, they will serve as coffer-dams, and being formed of the cheapest materials, their value as service-ground or platforms for the use of machinery, the moving of scows, &c. Their application to the sides of the piers is with particular reference to preventing the ice from reaching the spring of the arches, which will be the lowest and most exposed part of the superstructure, if wood be used.'

In the first design for the Victoria Bridge, 'ice-breakers,' very similar to the above-described by Mr. Keefer, were introduced; but subsequently the arrangement was changed, partly with a view of gaining the assistance of the whole weight of the bridge to resist the pressure of the ice (but it became fixed), and partly for the purpose of obviating the considerable annual outlay.

I have not data at hand to estimate correctly the cost of the ordinary 'ice-breakers,' as described; but I have little or no doubt that, as I before stated, they would have required to have been large and substantial masses of stone and timber, which in amount of cost would be scarcely less than, if not equal to, the inclined planes of masonry which have been added to the up-side of the piers.

It is now necessary for me to say a word or two upon the style of workmanship. It consists simply of solid ashlar; and considering the severe pressure and abrasion to which it will be subjected by the grinding of the ice, and the excessively low temperature to which it will be for months periodically exposed, I am confident that it is not executed with more solidity than prudence absolutely demands; and considering the difference of the rates of wages in Canada and this country, I believe the price of the work will come out nearly the same as any similar work let (here) by competition.

The description and style of the masonry is precisely similar to that adopted in the Britannia Bridge; the material is the same, and the facility of obtaining it is not in any important degree dissimilar.

The following is a brief description of this remarkable structure.

The site of the bridge is at the lowe end of a small lake, or enlargement of the stream, abo t a mile above

Montreal Harbour.   At this point the River St. Lawrence
is, from shore to shore, 8,660 feet, or nearly a mile
and three quarters wide.

The level of the river varies at different parts of the
year.   From about the middle of April to the end of
December it remains at what is called summer level, only
varying a foot or two above or below a certain line ; but
during the other $3\frac{1}{2}$ months, when it is covered with ice,
it rises about 10 feet higher, and at the beginning and
end of the period, sometimes as much as 15 feet higher.

The depth of the river is but shallow, varying from 5
to 15 feet below summer water level.

The current in the principal channels runs at the rate of
seven or eight miles an hour.

The bed of the river consists of slate rock, which lies
bare near the shore, but is covered towards the centre to
a depth of 12 or 14 feet with a deposit of clay and gravel,
so hard that it was at first mistaken for the rock itself.
Large boulders, varying in weight from 1 to 20 tons, lie
scattered profusely about, often appearing above the
summer water level ; the whole of this overlying matter
had to be cleared away at the site of the piers, so that
their foundations might rest upon the solid rock below.

The approaches are carried some distance into the
river on each side, partly in embankments and partly
in abutments of masonry, so that the length of the bridge
proper is about 6,650 feet.   This is divided by stone piers
into twenty-five openings, of which the middle one, serving
as the principal channel for the navigation, is 330 feet
wide, and the remainder are each 242 feet wide.

The bridge is constructed only for one line of railway,
the superstructure being a single iron tube extending

from end to end, through the interior of which the trains pass, in the same manner as in the Britannia Bridge.

The height of the bottom of the tube above summer water level is about 36 feet at the abutments, rising by a gradient of 1 in 130 to 60 feet at the centre opening.

Of the twenty-four piers of the bridge, all except the two middle ones are alike in horizontal dimensions, but they increase in height towards the centre, according to the gradient of the tubes.

Fig. 9 is a side elevation of one of the piers, and

FIG. 9.

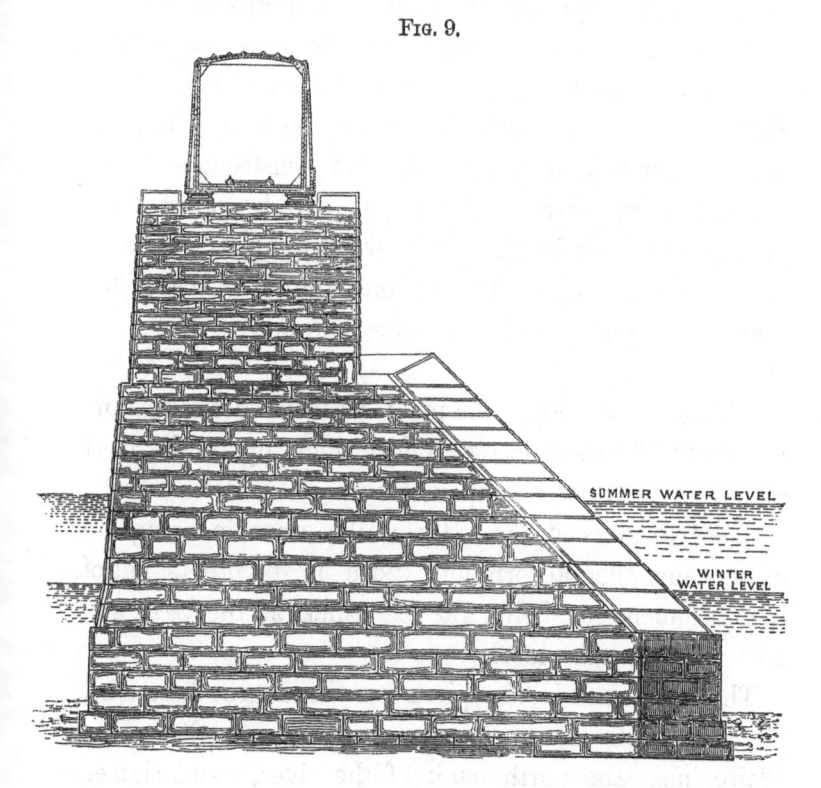

will give an idea of their construction. They are built of solid masonry, composed of heavy stones, from 5 to

20 tons each. They are founded with their footings resting on the solid rock, and are carried up, with the dimensions of about 90 feet wide by 23 feet thick, to within 6 feet of summer water level, at which point the ice-breaking plane begins. This is a slope of masonry, on the up-side of the pier, inclining backwards at an angle of about 45° with the horizontal, until it reaches a height of about 20 feet above the highest or winter level of the river. The face of this slope is pointed, like a cutwater, and the masonry is formed of large solid stones worked perfectly smooth, and strongly bound together internally with iron cramps, to resist the enormous thrust upon them. When the large floes, or sheets of ice moving down the river, come in contact with these massive constructions, they are turned upwards by the slopes, and breaking down or toppling over by their own weight on each side of the cutwater ridges, they fall into the open spaces between the piers, and so pass harmlessly down the river.

Above the ice slopes the piers are carried on, measuring about 33 feet wide by 16 feet thick, up to the level at which the tubes rest upon them.

The two piers at the sides of the centre opening are of the same general form and width as the others, but of larger dimensions, being 30 feet thick at the base, and 24 feet at the tube level.

The stone used in the piers is a hard grey limestone, obtained partly from quarries at a place called Pont Claire, near the north bank of the river, about sixteen miles above Montreal, and partly from an island in Lake Champlain, whence it was brought by the River Richelieu and the Champlain Railway. Although coming from points widely separated, these two kinds of stone are of very similar quality.

The quantity of masonry in the piers and abutments is about 2,713,000 cubic feet.

The tubes for the smaller spans are 16 feet wide, and 18 feet 6 inches high at the abutments, increased to 22 feet in the middle of the bridge. They are made of wrought iron plates, on a similar principle to those of the Britannia Bridge, except that the top and bottom are not cellular, but are formed simply of layers of plates riveted together, and stiffened by ribs, gussets, and T irons.

The centre tube has the same width as the others, but is 23 feet high, and the top and bottom have extra strength and stiffening.

The tubes are united in pairs, the middle of each double length being fixed firmly down upon a pier, while the two ends, resting on the two adjacent piers, are left free to slide upon rollers, to allow for expansion and contraction; a small interstice being left for this purpose between them and the next adjoining tubes at either extremity.

No advantage is, however, taken of the principle of continuity, each half being designed as if it were an independent beam.

A single line of railway, 5 feet 6 inches gauge, is laid on longitudinal sleepers in the middle of the tube, and a footway 4 feet wide is placed on one side of it for the passage of the railway servants.

The tubes are lighted by holes cut in the sides at every 60 feet, and they are protected by a light covering from the weather.

The weight of iron in the whole line of tube is 9,044 tons.

There were above three thousand workmen employed in the construction of the bridge, and the contractor's plant comprised four locomotive engines, together with six steamers and seventy-five barges, having a collective tonnage of 12,000 tons. Upwards of two millions and a quarter of cubic feet of timber were used in the dams, platforms, and other temporary works.

The carrying out of this great design in Canada, comprising the getting in of the foundations, the building of the piers, and the erection of the iron superstructure, was a work of no ordinary magnitude and responsibility. Mr. Stephenson gave occasional advice in this matter, and sent out working drawings, accompanied with complete instructions for the putting together of the tubes; but the credit for the successful accomplishment of this portion of the work is principally due to Mr. Ross and to Mr. Hodges, the engineer sent out by the contractors. The work occupied between six and seven years, and a sketch of the principal events it comprised will complete the notice of the structure.

The general plan of the works having been decided on, operations commenced in the winter of 1853. The river being frozen over, and having assumed its ordinary winter height, the ice was cleared and levelled along the line of the bridge; the sites of the piers were carefully set out, and permanently marked by iron pins driven down into the bed of the river; soundings were made in their immediate localities, and the most eligible channels for boats and barges were selected and defined.

The plan originally proposed for founding the piers was by means of large floating ring-dams of the nature of caissons. Each of these was to be large enough to encircle a whole pier; it was to be built on the shore, floated out to the site, moored and secured in position; then scuttled, sunk, and puddled, and the interior pumped dry to allow of the construction of the masonry. The only available time for operations in the river was during the summer months; and as no temporary works could be left in the river during the ice-season, the dams were so devised that, when the masonry within them was completed, they could be readily floated again, and taken to a place of safety, to be used for other piers in the ensuing spring.

Each caisson was 188 feet long and 90 feet wide externally, the internal chamber being 102 feet by 42; the front part or bow was made wedge-shaped, to stem the current, and the stern or hinder part was movable, to allow of its being taken away after the building of the pier.

Dams on a similar principle were to be used for the formation of the north abutment. Six of these were constructed on the shore during the winter of 1853; and being launched in May 1854, were towed up to the site of the abutment, where the actual works of the bridge were first commenced.

The dam was completed and laid dry, and the masonry was commenced in August 1854; the work had been raised 6 feet above summer water-level when the winter set in, and all further operations were necessarily suspended.

In June, 1854, the dam was got into position for the first pier; the masonry was begun in July, and the pier was finished before the end of the working season. The dam of the second pier was in place in July, and the masonry was erected to a height of 4 feet above summer level before the winter.

But now doubts began to be entertained whether the plan of founding by caisson-dams was the best possible. The dams when floating only drew 18 inches of water, but even with this light draught great difficulty was experienced in navigating the shallow rapid waters with so huge a mass; and still greater trouble was encountered in getting them into position. It was, therefore, decided that an attempt should be made to found the next piers by dams constructed of a species of open timber framing, called ' crib work,' and much used in Canada. Local contractors skilled in this kind of work were accordingly engaged, and though great difficulties were met with from the velocity of the stream, the dam to No. 5 pier was completed during the working season.

Accidental delays had prevented the dams of Nos. 1 and 2 piers from being removed before the ice began to form : it was hoped, by strongly protecting them, they might be enabled to stand through the winter; but this hope was futile, for on the 4th of January, 1855, a general and violent movement of the ice took place, which completely destroyed and carried them away. Other injury was done also to the abutment dam, and to the embanked approach ; but the permanent masonry of the two piers and the abutment stood perfectly well, as did also the new crib work dam of No. 5 pier.

During the ensuing winter, little was done beyond providing timber and quarrying stone. In the summer of 1855 the works were resumed; the north abutment was carried up to high-water level; the south abutment commenced; No. 2 pier finished; No. 5 pier commenced; and crib work dams made and fixed for three other piers.

At the end of 1856, seven piers had been finished at the north end, and two at the south end, and the masonry of the south abutment had been brought up to a high level.

In 1857 five more piers were erected on the south side, and two commenced on the north.

It was, however, now felt that the completion of the bridge was likely to be protracted for many years, owing to the shortness of the working season. The available time each year was, at the outside, six months; the earlier portion was occupied in preparing for the setting of stone, which could, therefore, seldom begin before the middle of August, and all mortar work ceased by the end of November, when the frosts set in; so that sixteen weeks constituted the whole working season for executing the masonry of the bridge. This fact induced, during the year 1857, the adoption of a very ingenious device, of bedding the ashlar masonry in felt instead of in mortar, as had been previously done with success at St. Anne's Bridge over the river Ottawa; and thus the erection of the masonry could be prosecuted during the winter. Strips of asphalted felt, about three inches in width, were laid along the whole of the front of the masonry, at such a distance from the edge that the work might be effectually pointed. On each of the cross joints similar

strips were placed, as likewise at the back of the stones.
As soon as one course of ashlar was laid, it was dressed
perfectly fair on the bed to a straight edge for the reception of another course, which was superimposed in a
similar manner, the backing being laid dry and packed as
closely as possible. Open spaces or flues were left, about
one foot square, throughout the whole height of the pier.
The work was completed in this manner during the
winter, and as soon as the weather permitted, and the
frost was fairly out of the stone, the piers were carefully
pointed and the whole of the interior well grouted from
the flues. The whole thus became one solid mass, the
clear water, which filtered through the joints, showing
very accurately the progress of the grouting. This admirable contrivance hastened considerably the completion
of the bridge.

During this year also was commenced the erection of
the iron superstructure, portions of which had been
already received. A timber staging or platform, of great
strength and stiffness, and supported at two intermediate
points by temporary piled piers, was fixed over the first
opening, and upon this the tube was erected. The whole of
the ironwork had been accurately manufactured and temporarily put together at the contractors' works at Birkenhead, under Mr. G. R. Stephenson's supervision, and every
piece was carefully punched and marked before it left
England, so as to define its proper place in the structure.
Owing to the accuracy with which this was done, the various
pieces, during the erection in Canada, were fitted and riveted
together without difficulty. The bottom of the tube was
first completed, and adjusted to level and camber; the sides
were next added, commencing at the centre, and as these

advanced towards each end, the plating of the top closely followed.

When the tube was finished, its supports were struck away, it was allowed to take its own bearing, and the piers and the temporary platform were removed for use in another place. The first tube was completed during the summer of 1857, and a platform was also fixed for the erection of the twenty-fifth tube, the first from the south end ; and the ironwork of this was also fixed during the winter of the same year.

In the next year, 1858, great progress was made, the number of finished piers being increased to twenty-one, and two others being brought up to above summer level. The remaining pier (the eleventh from the north side) was purposely delayed to give water-way for the navigation, until the large centre opening should be completed. Some modifications were made in the construction and arrangement of the dams, and some accidents and failures occurred, but strong efforts were made to push on the work, and, on the whole, the year's progress was very satisfactory.

In this year also five more tubes were erected on the north side, and five on the south side, while the platform was prepared for the large central tube.

In the beginning of the next year the erection of this tube was commenced, and continued day and night ; and it was finished and the supports removed by the end of March, only a few hours before the ice broke up. The most strenuous exertions were necessary to accomplish this, for it was evident that, if the general movement of the ice took place before the tube was clear of the temporary staging, it would risk the carrying away of the intermediate temporary piers, and the consequent entire destruction of the tube; and honourable testimony is borne to the energy and zeal with which every man concerned exerted himself, under great difficulties, and in the most inclement weather, to contribute to the desired result.

In May the dam was commenced for the last remaining pier, the eleventh from the north side; but a few days afterwards it was carried away by two large timber rafts which came floating down the river, and ran foul of this and one adjoining pier; many men, on the rafts and the bridge works, were thrown into the stream, but fortunately all were saved, though the destruction of property was considerable. A new dam was immediately commenced, and the pier was completed in September, and the fixing of the iron superstructure now proceeded rapidly.

In March 1859, Mr. Stephenson sent out Mr. B. P. Stockman, (who had taken an active part under Mr. G. R. Stephenson in the construction of the tubes,) accompanied by Mr. Samuel P. Bidder, the former traffic manager of the Railway Company, for the purpose of inspecting the progress of the iron superstructure; and a few months later, shortly before his death, Mr. Stephenson expressed a further wish that another visit should be paid to Canada, to examine and test the bridge on his behalf, previously to its being opened for traffic.

With this view, in November 1859, Mr. Stockman again went to Canada, accompanied by Mr. G. B. Bruce, who had been a former assistant of Mr. Stephenson's. Having examined the entire bridge, and carefully tested the

Drawn by S. Russell.

Engraved by H. Adlard.

Victoria Bridge over the St. Lawrence, Montreal.

tubes by running through them heavily loaded trains, they presented a report on December 17th, in which they recommended that, after a few small matters were finished, the Directors should accept the bridge from the hands of the Contractors as being completed satisfactorily. Mr. A. M. Ross, who had assisted in the experiments, concurred in the report ; the bridge was opened for public traffic two days afterwards, and the formal inauguration by H.R.H. the Prince of Wales, who visited Canada for the purpose, took place on August 25th, 1860.

The period of six years, which was occupied in the execution of this great work, seems by no means long,* when we consider the peculiar nature of the difficulties which had to be encountered, and which by the perseverance and energy of the persons engaged were successfully overcome. The extraordinary rapidity of the stream in summer, and the violent action of the ice in winter, were elements of a magnitude seldom entering into the operations of ordinary river engineering; and although the water was comparatively shallow, the nature of the bottom, and the immense boulders scattered over it, involved great difficulties. The rigour of the winter was also a heavy trial, not only in reducing the available working time, but also in its effect on the men engaged, who, for the most part, being new to the climate, were but ill prepared to brave its severity. In the summer the scorching heat struck them down by

---

* See some remarks by Mr. G. P. Bidder, contrasting the progress of their bridge with that of the Westminster bridge, Min. Inst. C. E. vol. xix. p. 227. It is right, however, to state that for the slow progress of the latter structure, the engineer, Mr. Thomas Page, was not responsible

coups de soleil; while in the winter they were frost-bitten and blinded by the glare of the snow ; the thermometer was often 50° below freezing point, and during the winter of 1858, when such exertion was made to get the centre tube finished, and when consequently night-work became necessary, the sufferings of the men were extreme. If there was any wind, the portions of the body exposed to it became instantly frozen, and the men had, therefore, to work in thick gloves and heavy coats; fur caps covered their ears, and heavy handkerchiefs were worn over the faces, leaving only a small portion free for vision. It often happened, when the wind blew up stream, that the men would become covered with icicles, and be obliged to leave their work. Notwithstanding all precautions, scores of men were frozen in their hands, feet, ears, and faces, and many had to go to hospital in consequence, but, fortunately, so prompt were the remedial arrangements that no serious consequences occurred. In the summer of 1854 the cholera made sad havoc among the men ; in some cases nearly a third of the number employed were sick at one time, and many of those who were not attacked ran away from the pestilence, so that the best men were lost to the work. To these misfortunes were added the difficulties of great general scarcity of labour ; the necessity of bringing out workmen at great expense from England ; frequent strikes and insubordination among the men ; and the discouragement caused by hostile parties among the inhabitants, who declared the attempt to build a bridge to be a defiance of Providence, and prognosticated its utter failure. In 1855 the cost of the works became so much increased by the general financial depression consequent on the Russian War, that the abandonment of the contract was seriously contemplated; and in a subsequent year great doubts were entertained whether the financial means of the Company would justify its continuance, nearly the whole of a valuable season being lost in consequence. Peculiar difficulties also arose from the isolation of the piers, for as there was no space at any of them for stacking or sorting the stones, every course had to be prepared at the quay, selected, and shipped upon barges, exactly in the order and at the time required. It followed, therefore, that a course, or often even a stone, wrongly sent, or a barge getting aground, or the loss or damage of any of the peculiarly shaped stones, many of which were of great size, caused the whole of the workmen, material, and plant, to remain idle till the want could be supplied, which often took a week or more.

However, all these difficulties were ultimately overcome, and the bridge remains a lasting monument, not only of the engineering knowledge and skill which designed it, but of the energy and perseverance of those who had to carry the design into execution.

W. P.

CHAPTER IX.

CONCLUDING YEARS AT HOME AND ABROAD.

Athenæum Club — Geographical Society — Royal Society Club — The 'Philosophical Club' of the Royal Society—Robert Stephenson, President of the Institution of Civil Engineers—Receives the honorary D.C.L. of Oxford—The Dark Side of his Prosperity—Failing Health—Admiration of Mechanical Skill—Speech at Sunderland—The Wear Bridge—Strong Affection for Newcastle—Periodic Visits to the Factory—Judicious and Considerate Conduct to humble Relations—Visits to Long Benton—Contribution to Painted Window in Long Benton Church—Visit to Wylam—Isaac Jackson, the Clockmaker—On Board the 'Titania'—Letter to Admiral Moorsom—Hampstead Churchyard—Social Engagements—Trip to Egypt—An unfulfilled Presentiment—Letters from Alexandria and Algiers—Last Christmas Dinner of Stephenson and Brunel—Last London Season—Last Visit to Royal Society Club—Last Will and Testament—Last Voyage to Norway—Opening of the Norwegian Railway—Banquet to Robert Stephenson at Christiania—Last Public Speech.

ÆTAT. 47-55.

ROBERT STEPHENSON'S clubs were the 'Athenæum' and the 'Carlton;' but the clubs in which he found society most adapted to his tastes were the periodical dinners of learned societies, or of coteries composed of certain members of learned societies.

He was a member of the Geographical Society, and he was a frequent attendant at the dinners of that learned body.

But the 'dinner-club' in which he most delighted was the Royal Society Club. Of the 'Philosophical Club' of the Royal Society he could not be a member, having neither contributed a paper to the transactions of any society, nor published a distinct treatise. To the last, the productions of his pen (with the exception of his article on ' Iron Bridges ') were official reports, or brief statements of fact, connected with professional operations. But the Royal Society Club, having no such exclusive condition attached to the honour of membership, on Thursday, April 26, 1855, the inventor of the Tubular Bridge was proposed for election by Professor Wheatstone, and seconded by Sir Roderick Murchison. At this period the club dined at the Freemasons' Tavern.

On March 6, 1856, Robert Stephenson made his first appearance as a member, Colonel Sabine being in the chair ; Sir John Rennie, Dr. Peter Roget, Professor Wheatstone, Mr. George Rennie, and Sir Benjamin Brodie being present.

At these dinners Robert Stephenson was one of the principal attractions and causes of enjoyment. He thoroughly enjoyed them, always stopping late for ' just another cigar and a little more talk '—and retiring at midnight to a friend's house, or another convenient club, for ' a little more talk and just another cigar.'

From 1856 to 1858, Mr. Stephenson occupied the Presidential Chair of the Institution of Civil Engineers. His inaugural address, on entering office, is printed in the Appendix.

On June 25, 1857, Robert Stephenson received the Honorary D.C.L. degree of Oxford, together with Sir Colin Campbell, G.C.B. (Lord Clyde), Earl Powis, Sir George Cornewall Lewis, Bart., Sir John M'Neill, K.C.B., Isambard Kingdom Brunel, and Dr. Livingstone (the explorer).

But, with all his social success, Robert Stephenson's life
had in these latter years much of sorrow.   He had reached
the period of life when men who have no children confess to themselves that the glory of their days is only a
shadow.   To those who enjoyed his inmost confidence,
he more than once revealed his sadness, and he was
counselled to rouse himself against despondency.

His health was irreparably broken; but to the last
he was so full of animation when in society that men
found it difficult to imagine him other than he appeared.
His hair had indeed turned white without long warning,
but it was remembered that George Stephenson had
a snowy head while he was still in the prime of man-
hood.   There were those also who could tell how the
amiable and gentle-tempered man began to manifest
a passing peevishness and irritability on trivial provoca-
tions.   Those who knew him thoroughly saw in these
and other symptoms the conclusive proofs of serious
mischief affecting health.   But few suspected how he
struggled against melancholy, and how he looked forward to death.   The quiet of his house, when it was
without guests, he could not endure.   Often he walked
about the lonely rooms, and sat down to yield to sorrow
which in the presence of others he courageously sup-
pressed.

In these last days he used to look regretfully on
the scenes of his early professional triumphs, and of his
wedded joy in the little house in Greenfield Place, Newcastle.   'The Robert Stephenson of Greenfield Place is
the Robert Stephenson I am most proud to think of!' he
once said to a lady.   He was at all times very fond of
the mechanical department of engineering, and to the
last no part of his cares afforded him more pleasure than

the direction of the Newcastle factory. His admiration of ' really good, honest mechanical labour ' was enthusiastic. If he railed paradoxically at new-fangled notions for educating workmen, he did so from a lively sense of the comparative worthlessness of superficial education. For the ' skill of artisans ' he had a strong poetic sympathy, and as his career drew to a close, his affectionate appreciation of the class from which his father had sprung manifested itself in many pleasant ways.

When he, in company with the members of the Institute of Mechanical Engineers, visited Sunderland in 1858, and received an address from the workmen then engaged in preserving that noble relic of Thomas Paine's genius for mechanics—the bridge over the Wear—he said in reply—

There are no members of society for whom I have a higher respect than for industrious and intelligent workmen. It is to them that the engineer is indebted for the full and efficient realisation of his conceptions—which, however good they may be, must largely depend upon the skill of the workman for their success. The progress made in the higher branches of engineering during the last thirty years, may be attributed, in a great degree, to the improved skill and intelligence of the workmen. The advance of mechanical science, and its application to useful purposes, must always go hand in hand with the skill and also with the comfort of the working classes. I cannot refer to a better example in proof of this than the bridge upon which we are now standing. The alterations and improvements which you are so admirably carrying on, could not have been executed at the time when the original bridge was designed. If the engineer, therefore, had even designed the bridge as it is now intended to be made, his mental labour would have been vain and useless, for there was not sufficient skilled labour in the country to realise such an idea. I merely take this bridge as an appropriate example on the present occasion, because it is a work you are now carrying on under my own direction ; but it is only necessary to look around, and we meet everywhere with engineering works to which the remarks I have just made apply in the strictest manner; and reflection on such subjects teaches us to feel that skilled labour is the great fulcrum upon which all our social progress depends, and that the success of this progress is just in proportion to the skill of the labour brought to bear upon the great works so thickly scattered thoughout the country.

The bridge over the Wear, of which Robert Stephenson spoke in these terms, consists of a single circular arch of 236 feet span, with a rise of 34 feet. As the springings commence at 95 feet above the bed of the river, the whole height of the structure above low water being about 100 feet, vessels of from 200 to 300 tons burthen can pass under it, without striking their masts. It was built by Thomas Wilson, at the instigation of Rowland Burdon, the Member of Parliament for the county; and the engineer, in accomplishing his task, carried out successfully Tom Payne's ideas with regard to open voussoirs. 'The stability of the bridge,' observes Robert Stephenson, in his article on ' Iron Bridges,' ' has been at all times, however, extremely precarious, and ordinary prudence cannot much longer delay its entire removal.' As another part of this work mentions in detail the operations of *extraordinary* prudence by which that wonderful arch has been made durable, it is unnecessary here to dwell upon them. It is enough to say that the labour of restoration, not entirely completed till Robert Stephenson was placed in the grave, was his *last* work— a work carried out by Mr. G. H. Phipps, the same engineer who had assisted him in his early labours on the locomotive.

Both by deed and word Robert Stephenson showed his

care for workmen—especially for the workmen of Newcastle. The Newcastle Philosophical and Literary Institute was embarrassed with a debt of £6,200, when, mindful
of the benefits he had derived in youth from its library
and lectures, he volunteered to pay off half its debts,
provided that the rest of the incumbrance was wiped off
by a public subscription, and that the subscription for
members was lowered from two guineas to one guinea
per annum. This latter condition was insisted on, for
the sake of the many workmen who, in Newcastle and
the immediate neighbourhood, are bent on the work of
self-education. From time immemorial, the working
classes of the Northumbrian coal-field have abounded
with George Stephensons—of less commanding genius
and less kind fortune.

In like manner, when it was proposed, in honour of
George Stephenson, to erect the Willington Memorial
Schools on the site of the house in which he dwelt, whilst
acting as brakesman to the ballast-engine, Robert Stephenson came forward with open purse. The Memorial
Schools, the Gothic architecture of which relieves the eye
of steam-boat passengers, grown weary of the wharves
and factories of the Tyne, were mainly raised by the
younger Stephenson's wealth. Of the £2,500 expended
upon them, £1,200 was his donation, £600 came from
his executors, the rest of the sum being made up by the
Government Educational Grant and a few private sub-
scriptions.

A visit to Newcastle, to look over the factory, was to
Robert in his later years an excursion of pleasure rather
than of business. Tyneside men were genuinely proud
of him. As soon as it was rumoured that he had

arrived at his hotel, friends hastened from all quarters
to 'the Chief;' and others, who could scarcely claim
the honour of his friendship, came on divers pretexts,
or with no excuse at all, to pay their respects to 'Mr.
Stephenson.' The morning after his arrival, when he
walked down to 'the works,' there was an unusual stir
in the thoroughfares, and the number of times the visitor-
townsman had to nod, or raise his hat, or shake hands,
was a strong testimony to the regard which all classes
entertained for him.

As he approached the factory, the old dog, that had for
years spent eleven hours out of every twelve in slumber,
roused himself and walked sedately up the lane, to lick his
boots, and receive a biscuit from his hand. A buzz
amongst the workmen testified how thoroughly the general
was beloved by the privates in his army. Amongst them
there were many of his near relations—first and second
cousins of the whole or half blood, and some few who, on
similarity of name, laid claim to a kinship that did not in
reality exist. Robert's conduct towards his almost count-
less poor relations, on his mother's side, deserves a word of
notice. He could not have raised them above the lowly
condition of their birth, had he desired to do so ; and
even if the will and the power to exalt them had been his,
he would have done no good by removing them from a way
of life in which they were useful and honest members of
society. But he never looked coldly on any of his kindred.
He was well pleased to know that they were *good workmen*;
and he was careful that his influence should tell in their
favour, without rousing hopes that could only in the long
run beget disappointment and discontent. There was con-

sequently an understanding, based on a healthy clannish sentiment, that workmen with 'the Chief's' blood in their veins were to be received at 'the works,' tried, employed, and advanced according to their merit. When trade was dull, and hands had to be put on short work or dismissed, kinship was a benefit, for 'the Chief's' relations, provided they were industrious and of good character, were protected against the reverses to which labour is liable in times of commercial stagnation. And when Robert was brought personally in contact with them, he conversed affectionately with them 'as relations,' inquiring after their common kindred, and reminding them of the time when he used to be entertained in their mothers' cabins.

In the autumn of 1857, Robert Stephenson stayed longer than usual at Newcastle, and made excursions in the neighbourhood to the familiar scenes of his youth. Accompanied by Mr. Matthew Bigge and Mr. Charles Manby, he went to Killingworth. He always enjoyed a trip to the old cottage on the West Moor, but this autumn the events of the jaunt were unusually gratifying.

After visiting the ship-building yard of Messrs. Mitchell and Son at Walker, the three companions went on to Long Benton. On their way they passed the blacksmith's shop to which Robert in his boyhood used to trudge with a load of picks on his shoulders. 'Ay,' he exclaimed, 'that's where I used to carry the pitmen's picks to get them mended.' Coming to the rivulet that runs under the bridge near Long Benton churchyard, he said, ' That's where I have fished for many an hour.' In the same way, on entering the 'parson's field' before the West Moor cottage, he observed that 'it was the field in which he used to torment the cows, by bringing down the electricity to their tails, by his kite-string.'

As they crossed the colliery tram-way before the cottage, they came on an old man, stationed there as gate-keeper.

'How long have you been here?' asked Robert, accosting the man.

'Why, aboon forty year,' was the answer.

'And what do they call you?'

'Why, they ca' me Clark, bot wha' ar' thoo?'

'I'm Robert Stephenson.'

'What! Robert Stephenson?' stammered the old man, collecting his wits.

'Yes, George's Robert,' answered Mr. Stephenson.

At the time these sentences were exchanged, the old man was getting his tea, holding a hunch of dark bread in one hand, and a tin can of hot tea in the other. Becoming greatly excited, the aged workman put down his tin vessel and bread on a bench, and grasping Robert by both hands, exclaimed, 'Eh man, but a's varry glad to see thoo.'

And then with tears in his dim eyes, a smile on his face, and a choking in his voice, the veteran went on—'Mony a time I've paid (i.e. beaten) thy heed, for thoo was a hemp (i.e. an idle, saucy fellow), and thoo was niver oot o' mischief, when thoo cam on th' pit heap wi' thy fether's meat.'

Passing on to the cottage, Robert inspected the sundial, the plan of which his boyish hands had traced; and then knocking at the door he asked for admission to the rooms in which so large a period of his early life had been passed. Of course he was received with hearty Northum-

brian welcome by the good woman who opened the
door.

The first thing that struck Robert's eye on entering the
cottage was that the little recess in the wall, where his
blackbird best liked to stand forty years before, had been
blocked up. 'What has thoo done with my blackbird's
corner?' was his enquiry to the surprised dame, who
marvelled not a little at the question. The next thing
that caught his eye was a piece of furniture, embracing
the conveniences of escritoire and book-case, that reached
from the floor to the ceiling of the principal room. It
was the work of his father's hands, and in the autumn of
1859 remained in the room where its artificer placed it.

'Do you know there is a secret drawer in that desk?'
Robert enquired of the dame who was playing the part
of hostess.

'It has nae secret drawer,' answered the woman
sturdily.

'Oh, but it has;' replied Robert, stepping forward as
he spoke, 'I know it has: for it was made by my
father.'

In another moment the button was touched, and the
hidden drawer flew open : but to the disappointment of
the spectators it was empty.

Till that moment the people of the cottage had been
ignorant of the concealed contrivance, and of the fact that
the piece of furniture was made by George Stephenson.

After gossiping with the people of the cottage for a
few minutes longer, Robert Stephenson rose with tears in
his eyes, and for the last time crossed the threshold of
the old home. Miss Tate (the niece of Robert's play-
mate, John Tate), who three years since, when collections

for this biography were being made, was hostess of the ale-house (the 'Closing Hill House') at the West Moor, recalls how Robert, in the year preceding this last visit to Killingworth, called at the ' Closing Hill House,' and enquired for the landlord, Robert Tate—John Tate's brother.

Robert was then suffering from illness, and so grey and changed in appearance had he become, that Robert Tate on entering the room did not recognise him.

' What, don't you know me, old friend ? ' asked Robert, much affected.

' Why —' said Tate, after a pause, ' it must be Robert Stephenson.'

' Ay, my lad,' answered Robert; ' it's all that's left of him.'

Then, sitting down in the little ale-house parlour, Robert Stephenson spent an hour and a half talking affectionately about days that were of the past to his ' old friend ' Tate, and to the neighbours who chanced to ' drop in.' Recalling the great man's demeanour on that occasion, Miss Tate says, ' He was full of condescension.'

Just about the same time that Robert paid his 'last visit to the old home,' he went over to Wylam, and looked once more at his father's birth-place.* Among

* Dr. John Besley, the vicar of Long Benton, communicated for this work a fact that pleasantly illustrates Robert Stephenson's affection for the parish of his boyhood. In 1855, when Long Benton Church was being rebuilt, Dr. Besley wrote to Tommy Rutter's old pupil, asking for a subscription to a painted window, and in reply received the following note :—

24 Great George Street,
Westminster,
March 24, 1855.

Sir,—I am glad to learn that Long Benton Church is being rebuilt, and as it is the only church with which I can associate my early boyhood, I beg to respond to your application for the painted window by enclosing ten pounds.—Yours truly,
Robert Stephenson.
Rev. John Besley.

other village worthies whom he then honoured with a call was Isaac Jackson—one of those ingenious, self-taught mechanicians, with whom the black-field of Northumbria abounds. Isaac Jackson's clocks are well reputed in pitmen's cabins for miles round, and at the time of Robert Stephenson's last visit to his father's native village, the foremost craftsman of the little community was busy in making a clock of more than ordinary excellence. Ever ready to show his sympathy with genius labouring under difficulties, Robert gave Isaac Jackson an order for a clock—made in his very best style. With due deliberation, Isaac executed the commission by the end of the following year. On December 14, 1858, the clock was sent to the late Mr. Weallens at the works, and when Robert received it, he not only paid the sum charged by the maker (£33), but added as a complimentary fee £6. One would like to know by what computation Robert restrained his liberality, so that the entire sum paid was just £1 short of even money.

In the same autumn, after visiting for the last time Gateshead and Killingworth, Robert Stephenson started on a yachting expedition with his friends Mr. Kell, Mr. G. P. Bidder, and Mr. Elliot. In a letter, which gives a picture of 'Life on board the Titania,' Mr. Kell, on his return to Gateshead, wrote to his sister :—

October 26, 1857.

MY DEAR SISTER,—I have had a most delightful excursion since I saw you at Harrogate. My old friend and schoolfellow Robert Stephenson came down here in his schooner yacht, 'Titania,' with a crew of sixteen men, a good cook, and a first-rate cellar—and he impressed me on board on a voyage to Aberdeen, Peterhead, Cromarty, Inverness, along the Caledonian Canal, through the magnificent Lochs Ness, Aich, and Lochy to

Loch Eil. We had an ample supply of astronomic and mathematical instruments; and one person on board, at least, knew how to use them. We made repeated observations on the temperature of the water at the surface, and at various depths, at one place in Loch Ness at a depth of 170 fathoms. Mr. Bidder, Mr. Elliot, and I composed the guests, and our discussions over our cigars in the evenings were most interesting. * * * * Then we bore away for Holyhead; and having examined the gigantic works of the harbour of refuge there, we devoted a day to the Britannia Bridge. I can never forget the interest which the designer and executor of that magnificent monument of skill and enterprise excited in us, as he described in his quiet way the general design, the objects to be effected by the different parts, the difficulties encountered and overcome in the erection, and the fact that if each of the enormous tubes were sawn through in the middle, the bridge would carry the trains. The principal part of the description was given on the top of the tube, on a beautiful morning, in full view of the Naples-like scenery of the Menai Straits, and the distant Welsh mountains, Snowdon and its associates. We smoked a cigar in quiet contemplation before we left the spot, none of the party being disposed to speak, and returned to Holyhead quite delighted. The next morning I had intended returning home, but when I got on deck, the yacht was spanking before the wind, which in a very few hours wafted us to Kingston Harbour. Judge Keogh gave us some amusing reminiscences of his discussions in the smoking-room, with Bright, Cobden, Stephenson, and other friends of both sides of the house. * * * The judge, his lady, and a party paid us a return visit on board the 'Titania,' and were delighted with the elegance and accommodation of the saloon and cabins, but not particularly so with the boat passage in the harbour, where a rolling swell had been raised by the easterly wind, which had begun to blow, and which, increasing to a gale, has since caused fearful damage and loss of life. The only misfortune that befell us was on the night we sailed from Sunderland, when, off the Bell Rock, the wind freshened and northered, the sea rose, and at 11.30 P.M. two waves (sailors call them *seas*) met on our deck, and broke on board—filled the cutter which was on davits seven feet above the deck, and broke the after davit (an iron bar of great strength). The boat was recovered with great difficulty, *minus* the oars, benches, and floor. Mr. Stephenson was standing on the weatherside of the saloon. He was pitched off his feet, and thrown to the lee side, with his head against a lamp fixed by a gimbal to the waintscoting, breaking the lamp, and, the bronze gimbal cutting his head severely. He was also bruised by falling against the front of the sofa. I was in bed, and tried to get up, but could not keep my feet until the vessel was brought up to the wind. Mr. Stephenson suffered from the bruises for a few days, but recovered ere we got through the Caledonian Canal, and we were as hearty as crickets. I have told you what we did. I must now tell you how we lived. There was a capital library on board, and a gimbal lamp at each bed-head, and each man before going to bed selected a book. At seven in the morning a cup of coffee was served in bed to each man, who then read or snoozed till nine; when the decks having been washed, the brass hand-rails and passages all cleaned, he dressed and came on deck.

The good library of the ' Titania ' is a fact worthy of notice. In years intervening between 1850 and 1859, Robert Stephenson, having more leisure, became a more general reader than he had been. He read books of all kinds, and on a great variety of subjects—selecting them himself, and judging of them for himself, without the aid of critical guidance.

The parliamentary season of 1858, Robert Stephenson passed principally in town—apparently enjoying average health, but really giving way to confirmed disease. Quite reconciled to the thought that his life would end in the course of a few years, he maintained his old cheerfulness of demeanour in society, and even in solitude he was less subject to fits of melancholy. But the solemn reflections induced by his condition were not stifled or avoided. He began to take his horse exercise in the country, about

Hampstead and Wormwood Scrubs, instead of the parks, and occasionally he took solitary drives in his carriage. After his death, it was found that not seldom these drives took him to the churchyard in which his wife lay buried. It was near Mrs. Stephenson's grave that the strong man, as his strength failed him, could best meditate on the coming change.

His mind, however, was still intent on great works. Though he had withdrawn from the turmoil of his profession, he was inspiring and regulating the labours of younger or stronger engineers. In Norway, under his friendly surveillance, Mr. Bidder was finishing the Norwegian railway. His counsel aided Mr. Rouse in the construction of the Kaffr Zeyat viaduct over the Nile. In Canada, Mr. Ross was carrying to completion the grand viaduct which the inventor of tubular structures designed in the outset, and controlled in every perilous crisis. And in Sunderland, Mr. Phipps was acting upon his instructions for the preservation of the Wear Bridge. He was also deeply interested in the construction of Brunel's Leviathan Ship.

His appearances at the Royal Society Club, and his speech on the Suez Canal in the House of Commons, have been already mentioned. He went much into general society, and entertained his friends at the customary Sunday lunches. It was also the second year of his Presidency over the Institution of Civil Engineers, and whilst he filled that office he entertained his professional comrades at weekly dinners.

In the autumn he started for Egypt in the 'Titania,' accompanied by Mr. and Mrs. Perry, Captain Bedford Pim, and Miss Bidder. The 'Titania' left her moorings

in the Southampton Water at five o'clock A. M., Thursday
October 14, 1858, and arrived at Gibraltar on the 27th
of the same month.   After touching at Malaga, Grenada,
and Algiers, the party went on to Malta.   The passage
from Algiers to Malta was made eventful by a sudden
and terrific hurricane, in which the yacht and all on
board were within an ace of being lost.   On Tuesday,
November 3, the 'Titania' anchored opposite the Pasha's
palace at Alexandria.   On Friday, the 27th, the travellers
were in Cairo.

In Egypt, escorting a party of ladies to the antiquities,
Robert Stephenson was a happy man.   He had carefully
read every authority on the history, geography, and
natural features of the country.   The explorations of
antiquaries on the banks of the Nile, and disquisitions of
critical scholars upon them, were familiar to him as the
moves of the board to a chess-player, and he spared no
pains in communicating to others the results of his own
careful study.   Well qualified to be their instructor, he
exerted himself to give Mrs. Perry, and the other ladies
who took part in the excursions from Alexandria and
Cairo, clear and accurate views on every object presented
to their curiosity.

Writing to Mr. Thomas Longridge Gooch (then resid-
ing at Nice) during this expedition, he said :—

Alexandria : December 5, 1858.

DEAR GOOCH,—I was pleased to receive your note on my
arrival, but I find my friends have been so kind in writing that
I am overwhelmed at the work I have before me in replying to
them.   I must therefore cut you short with a very brief epistle.
Our voyage was, upon the whole, remarkably fine, rather too
much so, as the wind was generally light, with occasional calms

of two or three days' duration. Thus you will see that our
climate has been very different from yours. Indeed, I read the
papers with much surprise, as regards the lowness of the
temperature you had experienced at so early a period of the
winter. I heard also that Nice had suffered in a similar way,
although not of course to the same extent. Constantinople also,
I hear from a gentleman just arrived from thence, is really
miserable for cold and snow. They have a theory here which
I think is probably correct : viz.—that, when the north of Europe
suffers from severe weather, the countries adjacent to the tropics
have invariably fine seasons. Last year confirmed the notion,
and this year certainly does the same. We leave to-morrow
for Cairo, whence a portion of our party will most likely go up
to Thebes ; but having once been there I do not mean to go
again. I shall stay quietly at Cairo and enjoy daily a drive
into the desert, which I have always found most invigorating.
By the last post I had a message from Brunel, inviting me to
dine with him at Cairo on Christmas-day.

This I shall endeavour to do, although at some inconvenience.
This circumstance reminds me that I must wish you and your
family a merry Christmas and a very happy new year.

Believe me, yours sincerely,
R. S.

The climate and life of Egypt suited Robert Stephenson,
literally ' intoxicating him with delight.'

The two following letters, one written during his Egyptian excursion in 1856, and the other penned at Algiers
in February 1857, are pervaded by a cheerful tone :—

Alexandria : December 22, 1856.

Dear Clark,—Many thanks for your kind letter, and I am
glad to hear that my coat was of service to you on your way
home. You need not forward it to Alexandria, as I shall in all
likelihood have left before the next post arrives ; but, in case
you have done so, I will give instructions to Rouse about it.

I shall not here go into any particulars of our voyage, as I
have done so in my letter to G. R. S., which must be considered

common property at 24.* From it you will perceive that, in consequence of the Viceroy's absence, I shall probably take a run over to Constantinople, and return to Alexandria, if the intelligence I receive there holds out any prospect of my having an interview with his Highness after his return from Abyssinia— if he return at all; for I expect, if he persevere, he will be seized with the fever of the country, and never more be heard of.

With respect to your proposed visit to Malta, I can scarcely advise you to do so with the expectation of meeting me, but I am sure it will do Bidder a world of good. The climate is perfectly delicious, and if you only come to Malta for a week or ten days, it will do you both much good; and with such a stay at Malta, it is quite possible I may drop in upon you, in which case we could return by yacht to Marseilles.

My own health is quite a different thing here. I am quite in good spirits, without an atom of hypochondriac feeling, and actually recovering my flesh. A happy new year to you all!

Yours sincerely,

ROB. STEPHENSON.

In the following year Robert wrote to the same friend :—

Algiers: Feb. 19, 1857.

DEAR CLARK,—From Bidder you will learn of my movements up to this place, and my remarks on the present condition of the capital of Sicily, which I believe you saw years ago.

This place presents a very remarkable contrast with Palermo at the present moment. The population are active and cheerful, and commerce seems carried on with alacrity and success. The surrounding country is being rapidly brought into a high state of cultivation. Many wealthy French and English farmers (that is to say, *for* farmers) are extending the science of cultivation very rapidly, and, I understand from one of themselves, with perfect success and ample profit.

The old town of Algiers is of course a complete specimen of Moorish arrangements for domestic establishments—narrow

* At No. 24 Great George Street, not only Robert Stephenson, but several of his most valued professional friends—such as Mr. Bidder, Mr. Edwin Clark, Mr. Manby, Mr. Phipps, &c.—had their offices.

streets not more than nine or ten feet wide, with the upper
stories of the houses gradually projecting one beyond the other,
so that, to go from the upper rooms of one house into the upper
rooms of the opposite one, it is unnecessary to go downstairs.
The new town, built since the French took possession, may
very fairly be compared to some of the best portions of Paris—
excellent cafés and restaurants, elegant shops with all kinds
of gay merchandise, and *all other matters* which an advanced
state of civilisation calls for.

We visited a *café chantant* last evening, and it was curious
to see how quietly and gracefully the Arab, with his loose flowing
garments, mingled with and adapted himself to the European
people and their customs.   The contrast between the two
races in the coffee-room was not more remarkable and interesting than the contrast between the old town built by their
ancestors and the new town now being built by the French—
the one stolid and sombre, the other excitable and gay.   I
recollect that the locomotive in Egypt seemed scarcely to excite
an emotion in the Arab when he saw it rapidly moving along
with an enormous load.   Such a scene was entirely beyond his
comprehension—that is, his mental powers were not equal to
reflecting and reasoning upon it; but the moment he heard the
steam whistle, which touched an external sense, he was excited
and confounded.   This is precisely what we ought to expect
where the mind has undergone no culture.   The cultivated
mind appreciates every new phenomenon with interest and
surprise; this is a refinement beyond the reason of the savage.
The latter would see the crab creep out of his shell almost
as a matter of course, but such an event excites wonder and
confounds the philosopher for a time, till reflection and comparison with parallel phenomena convert the event of the crab
creeping out of its shell into another link in his chain of
reasoning. . . .

> Yours sincerely,
> ROB. STEPHENSON.

The arrival of Mr. and Mrs. Brunel at Cairo, on
December 20, 1858, gave Robert Stephenson great satisfaction.   Although he was greatly out of health, Brunel

was in good spirits, and, in the excitement of holiday-making, was ready to take an imprudent amount of exercise. When he became fatigued with walking, he mounted a donkey and rode about the streets of Cairo, to all appearance as free from care as a schoolboy. Robert had taken up his quarters at Shepherd's Hotel; Brunel stayed at the Hôtel d'Orient. At Shepherd's also were Lord and Lady Dufferin. Lady Dufferin pressed Robert Stephenson and his friends to dine with her on Christmas-day; but the prior engagement with Mr. and Mrs. Brunel precluded them from accepting the invitation.

On December 25, 1858, Robert Stephenson and Isambard Kingdom Brunel—the two greatest engineers of the nineteenth century—dined together, in company with a few mutual friends, at the Hôtel d'Orient, Cairo.

On his return from Egypt, Robert Stephenson stopped for a few weeks in Paris, returning to London on or about February 9, 1859. The change had greatly benefited him. The deep-seated mischief in liver, stomach, and nerves, of course remained untouched, but its most distressing symptoms were less apparent. He was cheerful, and enjoyed an unaccustomed sense of vigour. With characteristic ardour, he returned to the pleasures of English society and the duties of his position. The St. Lawrence and the Sunderland Bridges, and the Norwegian Railway (soon to be completed), the state of the Serpentine, and the Metropolitan Drainage were amongst the objects of his care. He was regular in his attendance in the House. His drives in the country were frequently repeated. In Gloucester Square he was as hospitable an entertainer as in former seasons.

He became very much interested in the proposition for an Atlantic Telegraph. From the day when the first telegraph was put down on his line between Euston Square and Camden Town, he had been a zealous promoter of telegraphic communication; and now, in the last year of his life, he consented to act in a commission * (composed of himself, Captain Douglas Galton, R.E., and Professor Wheatstone) appointed by the Lords of the Committee of the Privy Council for Trade, and the Atlantic Telegraph Company, 'To inquire into the Construction of Submarine Telegraph Cables.' The first meetings of the committee took place at his private residence. He was also chairman of the Electric Telegraph Company; and to his wise provision of a sinking fund to meet the expense of renewals, the prosperity of the company is mainly due.

Note-books and letters of chat show that he was continually in society. After dinners about town, he several times called on Dr. Percy, at Craven Hill, Bayswater, and smoked a cigar.

On April 11th he attended the meeting of the Geographical Society, where Captain Pim read a paper on the Suez Canal question.

The next day, Mr. Joseph Bonomi read a paper 'On the Means suggested by Robert Stephenson, Esq., M.P., for the Extraction of a ponderous Granite Sarcophagus

---

* After Robert Stephenson's death, other gentlemen were introduced into the committee. The report of the committee, when presented, contained the following words:—'In the first place, however, we must express to your Lordships how great was the loss we experienced, soon after the commencement of this enquiry, by the death of Mr. Robert Stephenson. His philosophical mind, his high scientific attainments, and his great practical knowledge, peculiarly fitted him for directing an enquiry such as this, in which mechanical, chemical, and electrical science are combined; it was a subject, moreover, to which he had given considerable attention, and in which he took the greatest interest.'

out of the Limestone Cavity in which it had been placed
by the Ancient Egyptians.' In the subsequent discussion,
Robert Stephenson, Mr. Sopwith, Mr. Perry, Mr. Jennings,
Mr. Sharpe, and others, took part.   Mr. William F.
Ainsworth (Honorary Secretary of the Syro-Egyptian
Society) recalls how all the members on this occasion
were struck by the modesty of Robert Stephenson's lan-
guage and tone.

On June 4th he was present at the annual inspection of
the Royal Observatory at Greenwich, and at the subsequent dinner at the Ship Hotel.   But as the season wore
on, it became manifest that his health was in a more
precarious condition than it had ever been.   With total
loss of appetite, and powers so deranged that the palate
could no longer distinguish flavours, with constant lassitude
and overwhelming weight of depression, the sick man
struggled on bravely.

On June 17th Dr. Percy called upon him in Gloucester Square, and found him, at half-past eleven A.M.,
eating a few strawberries.   On this occasion he spoke
more fully than he was wont of his wretched condition,
but looked forward hopefully to the time when he could
get away from town, and once again enjoy the sea-air.
But he did not permit his sufferings to depress him in
society.   A passage in the records of the Royal Society
Club * gives the reader a glimpse of him in these last
days : —

Among the more recent dinner-parties, that of August 11,
1859, may be noted, a curious incident in its components
having given, as will be presently seen, an unusual prepon-

* 'Sketch of the Rise and Progress of the Royal Society Club.' Printed
for private circulation.

derance to the delegates of practical knowledge; while the various walks of general and abstract science were also ably represented on the occasion. . . . .

Among the visitors on that day was Mr. Thomas Maclear (now Sir Thomas), the Astronomer Royal at the Cape of Good Hope, who had just arrived in England from the southern hemisphere after an absence of a quarter of a century, during which period, besides assiduous attention to his regular observatorial duties, he had measured an important degree of the meridian in Caffraria. . . . .

This was the last time that Mr. Stephenson, the celebrated civil engineer, attended the Club. He was not looking well, nor was he animated with his usual flow of cheerfulness; and he left the room early in order to take his seat at a debate in the House of Commons on cleansing the Serpentine. It was remarked that on this day were present, so to speak, the representatives of the three great applications by which the present age is distinguished—namely, of Railways, Mr. Stephenson; of the Electric Telegraph, Mr. Wheatstone; and of the Penny Post, Mr. Rowland Hill—an assembly never again to occur.

Before the next celebration of the anniversary of the Royal Society Club (July 5, 1860), death had been busy with its members. Dr. Percy had to announce the deaths of Robert Stephenson, General Leake, Rev. Baden Powell, and Robert Edwards Broughton.

It would have been strange had Robert Stephenson manifested ' his usual flow of cheerfulness.' As he sat at that dinner, the gloom of fast-approaching death overshadowed him. For weeks he had been getting worse in health, and he was convinced that before many more weeks had passed he would be in his grave.

The Club dinner was on August 11. On the morning of the 15th of the same month he embarked in the ' Titania ' for Norway, in order to be present at a banquet to be given him at Christiania on the opening of the Norwegian Railway, and in the hope that change and sea-air might afford him relief.

But, before leaving his native land, he finally revised his will. When it came to be proved, the personalty was sworn under £400,000. The executors appointed under the will were Mr. Charles Parker (the testator's solicitor), Mr. George Robert Stephenson, C.E., and Mr. George Parker Bidder, C.E. To his cousin, the above-mentioned Mr. George Robert Stephenson, the testator bequeathed all his interest in the locomotive steam factory at Newcastle, and his interest in the Snibstone collieries in Leicestershire, left him by his father; he also bequeathed to his cousin his leasehold house in Gloucester Square, with its furniture, pictures, statuary, plate, library, wine, and other effects, as well as half the furniture and effects in the office in Great George Street, and a legacy of £50,000. To Mr. George P. Bidder he left the other half of the office furniture and effects, and £10,000, a like sum being bequeathed to Mr. Charles Parker. The testator left also to his cousins, Robert and James Stephenson, £5,000 each; to ten female cousins, on his father's side, £1,000 each; to Mr. W. Weallens (his partner in the Newcastle Factory), Mr. G. H. Phipps, Mr. Edwin Clark, Mr. T. E. Harrison, Mr. W. H. Budden, Mr. P. H. Stanton, Mr. James Berkeley, Mr. George Berkeley, and Mr. W. Kell, £2,000 each; to Mr. James Green and Mr. Stockman, £1,000 each; to Mr. George Vaughan, £5,000; to Miss Emily Lister, £4,000; to each of that lady's two sisters, £1,500; for the support of the children of the late Mr. Starbuck, £5,500; to Margaret Tomlinson, his housekeeper, £100 per annum for life; to each of his other servants who had been with him a twelvemonth at the time

of his death, £20 ; to the Newcastle Infirmary, £10,000 ;
to the Newcastle Literary and Philosophical Institution,
£7,000 ; to the North of England Mining Institution,
£2,000; to the Institution of Civil Engineers, £2,000; to
the Society for Promoting Christian Knowledge, £2,000 ;
and to the Society for Providing Additional Curates in
Populous Places, £2,000. The residue of his estate the
testator left to his three executors, share and share alike.

Accompanied by a rather numerous party of friends,
Robert Stephenson went down to Harwich, off which port
the ' Titania ' and Mr. Bidder's yacht the ' Mayfly ' were
lying. Mr. and Mrs. Bidder, Mr. James Harby, jun., and
Mr. and Mrs. Perry went on board the ' Mayfly.' Robert
Stephenson had for companions in the ' Titania,' Mr.
T. L. Gooch, Mr. James Berkeley, Mr. Phipps, and Mr.
Haly. Between four and five o'clock A.M. on the 16th
the yachts got out to sea, entering the Christiania fjord
at 6 A.M. on Sunday, the 21st.

Landing at Christiania, Robert Stephenson was well
enough to enjoy a trip along the line from Christiania to
the other terminus, Eidswold. The party visited two
farms belonging to Mr. Bidder, in the vicinity of the line ;
and on September 2 they returned to Christiania, taking
up their comfortable quarters on board the yachts.

The next day (September 3) a grand dinner was given to
Robert Stephenson. The desire to do him honour may
be estimated in some degree by the fact that one hundred
and thirty persons were present at the entertainment,
who paid two guineas each for their tickets—a sum of
no trifling importance in Christiania.

The toast of the evening was introduced with a speech
concluding thus :—

It is sufficient to call to mind how highly our country and our city, which are so greatly interested in this railway, are indebted to Mr. Robert Stephenson. On behalf of our country, our King has already made him deserved acknowledgment. The Olaff Cross on his breast is the witness thereof. But we— the inhabitants of Christiania—what can we do? Little, very little, where the question concerns a man so significantly honoured in his own country—that mighty empire, over whose dominions the sun never sets! Can we add a fresh laurel to the wreath of honour that already crowns him? We can only offer him our modest ' forget me not,' and beg him for our own sake to keep it amongst those many and dear tokens of remembrance which his active life has procured him.

Therefore we have met to-day—men of Christiania, of all classes—not to try to honour Mr. Robert Stephenson, but to do honour to ourselves, by showing that we can esteem and gratefully acknowledge high merit.

Mr. Robert Stephenson, we offer you our most hearty thanks for your active zeal and disinterested service in a cause important to our country in general and our city in particular. We wish you health and happiness, and peace in your life's evening, which—and may the Almighty hear our prayers!—we trust may be bright and gentle—bright as the evening sun resting on snow-clad mountains, gentle as a quiet summer evening in our peaceful valleys.

It may perhaps be the last time you visit our country and its capital. We do not beg you to preserve in your mind our *country*, with its great mountains, clear lakes, and sunny vales. Nature herself has put so peculiar a stamp on our fatherland, and adorned it with such rich and lovely colours, that no one can see it and easily forget it. You, Mr. Stephenson, with a mind and heart open to nature's beauties, will be the very last to forget it, struck as you were by the first sight of it.

What we will only beg of you is, that, when you have returned to your active life in your own mighty country, you will allow your thoughts to take flight to the *capital* of Norway and to its grateful inhabitants.

Robert Stephenson had intended entertaining the assembly with a sketch of the rise and progress of railway enterprise—feeling that, though the majority of his audience would not understand the English tongue, they would appreciate his remarks when they read them on the following morning translated in their journals.   On taking his seat, he was so comparatively easy and free from pain, that he thought he should be able to carry out his intention.   But, just before he was required to rise, an attack of nausea and faintness rendered him powerless to do more than give utterance to the following remarks, which are of interest as being his 'last public speech :'—

Gentlemen : Although I was not able to understand the last speech, yet I heard my own name mentioned so repeatedly, that I know it was addressed to me. I have also had the honour to receive here a translation of its contents. I see from it that far greater merits are attributed to me than I should dare to attribute to myself. It is unnecessary to recapitulate the advantages of a railway to a country in advancing commerce and stimulating industry; they are truths everywhere acknowledged and well understood. But there are a few points in the last speech to which I will take the liberty of calling your attention. I find that the honour of the railway is altogether ascribed to me. This is not just. At the same time that you have yourselves to thank for it, there are also two other men to whom the honour is to be ascribed quite as much as to me, and indeed more than to me. I name first Mr. Consul-General Crowe, who, attentive to the capabilities of your country, conceived the idea, and gave the first impulse to the undertaking; and next Mr. Bidder, who has the undivided merit of carrying out the idea. He has the honour of having built the railway for an extremely low cost; he has built it for the sum of £450,000, which is less than, under the circumstances, could have been hoped or expected. It is thus to Mr. Bidder that the chief honour belongs for establishing this railway, which is now completed with English skill, aided by no inconsiderable Norwegian capital and liberality. It is quite true, as I read in the trans-

lation before me, that I have been occupied with great works
in other places.   I have been employed in Canada, in Egypt,
in Belgium, in Russia, and I may say in nearly every country
in Europe.   But if I could ascribe to myself the whole merit
here among you, I should act unjustly towards the two gentlemen
I have already named, Mr. Bidder and Mr. Crowe, to whom so
much is owing—to whom more is due than to me.   But let
me express my sincere gratitude for the present festival by
which I am so much honoured.   It is probably the last time I
shall meet the citizens of Christiania.   I shall leave your
country on Monday.   But I cannot leave you without expressing my best wishes for this town, my most cordial wishes for
Christiania, which has contributed so much to the existence of
this railway, and to whose prosperity and happiness this railway
will contribute so much.   I should wish once to come back
here to see the results—the advantages on which I now con-
gratulate the town.   Prosperity and happiness to Christiania!

Thus modestly did Robert Stephenson, in his last public
speech, direct to others the eulogies showered upon himself.
Throughout his career—from the day, when as a stripling
he directed the Colombian mines, to that occasion of
sad festivity at Christiania—was he thus cautious not to
appropriate to himself the honour due to his companions.

CHAPTER X.

LAST SCENES.

Homeward Course of the 'Titania' and the 'Mayfly'—Robert Stephenson lands at Lowestoft—Arrives at Gloucester Square again—Temporary Rally—Death—Public Agitation—The Queen's Expression of Sympathy—Funeral Procession—Interment in Westminster Abbey—Attendance at the Ceremony—Sacred Service at Newcastle—Public Mourning at various important Towns—Plate on Coffin-lid—Inscription on Monumental Brass—The Article in the 'Times' on the Morning after the Funeral—Generous Tone of the Press—Last Honours.

(ÆTAT. 55.)

THE next morning (Sunday) he was decidedly worse, and by night it was felt that, unless he made good speed to England, he would most probably die away from his native country. Early on Monday morning (September 5) the yachts dropped down the Christiania fiord. When they were just off Dröbak—from which place the London market obtains the ice sold as ' Wenham Lake Ice '— he became much worse, and there were good grounds for fear that he would not reach the British coast alive. While the passengers on board the ' Mayfly ' were at breakfast, Mr. Gooch came alongside in one of the ' Titania's ' boats, begging Mr. Perry (who was a member of the medical profession, though he had relinquished practice) to pay ' the chief ' a visit. The wind was up, and momentarily blowing stronger, and the yachts were fast approaching the open sea, when, in case of rougher weather, it would be impossible to send messengers from one vessel to the other. It was therefore decided that Mr. Gooch should take back with him to the 'Titania' Mr. and Mrs. Perry, and replace them without delay by Mr. Phipps and Mr. Haly. This shifting of passengers would give the invalid the benefit of medical attendance.

Two hours later, and the change could not have been effected. Heavy weather ensued, and the homeward passage was rude and perilous enough to try the endurance and fortitude of trained sailors. Jaundice, in a most aggravated form, had supervened upon other mischief, and the dying 'chief' passed days and nights in extreme danger. To make matters even more dismal, the 'Titania' was denied the companionship of the 'Mayfly.' In the darkness the two yachts missed each other. The 'Titania' being by far the slower craft, her captain, judging that the 'Mayfly' had unawares gone much ahead of her, deemed it best to make all possible way, in the hope of overtaking her. On the other hand, the 'Mayfly,' knowing her own superior powers, slackened pace so that she might be overtaken, when, in reality, instead of having outstripped the 'Titania,' she was far behind her. The consequence was, that the vessels did not again speak with each other until the close of the passage.

On the 13th the 'Titania' neared the Suffolk coast, and after beating about for several hours in darkness, she found the 'Mayfly' close beside her, waiting to enter Lowestoft port. As no arrangement had been made for the two parties to meet at that point, the occurrence was an agreeable surprise to both. A pilot having been obtained, the yachts entered the fine harbour of Lowestoft at early dawn. The sight of 'the old country' did the sick man good. He was well enough to walk on shore from his yacht to the railway station, where he had to wait a short time for a train. It was suggested that he should be carried, but he insisted on walking—and he walked, leaning on the arms of two friends.

To convey him from the coast to London was the work of a short time. The sight of home appeared to restore him, after the fatigue and suffering of the preceding days. No time was lost in sending a messenger for Dr. Frederick Bird, who with Dr. Baly remained in constant attendance until the end.

From the day of reaching Gloucester Square till the following Sunday he seemed to improve. Certainly he gained strength. On Sunday afternoon, without medical permission, he astonished the members of his household by appearing in the drawing-room, and declaring that he had grown tired of his bedroom, and was resolved no longer to be treated as a sick man. But the exertion and the excitement of conversation were too much for him, and he was carried, rather than led, back to the apartment in which he died.

The next morning he was much worse; obstinate congestion of the liver was followed by dropsy of the whole system, and it was seen by his physicians that the close was not far distant. Aware of his position, he manifested a serenity that would have been perfect, had it not been occasionally broken by anxiety for Mrs. Bidder and the other affectionate friends who watched him day and night. For death he had no fear. Indeed, as a place of rest the grave he had long been prepared for.

Whilst death stood over him, the concern of the public

mind was deep and universal. A continual stream of callers enquired at the door, and intelligence of his state was daily sent by rail or telegraph, to the chief centres of British industry.

At mid-day (shortly before the clocks marked the hour of twelve), on October 12, 1859, Robert Stephenson breathed his last.

As soon as it was learnt that he was no more, there was an universal exclamation of regret. Although his celebrity had made him a familiar name in every civilised community, he was, in the ordinary sense of the term, ' a private person.' It was only occasionally that he had acted as the commissioned servant of the entire nation. Never before had the death of a private person struck so deeply the feelings of his country. Coming so soon after that of the younger Brunel, his irreparable loss left a blank which no one living engineer could be expected to fill.

The regret was uniform throughout all ranks. Neither party nor clique contradicted the one prevailing sentiment. The man who in his life had been incapable of jealousy or meanness, was followed to the grave by men of every shade of opinion.

Not five days after his death, the following letter was penned by the Viceroy of Egypt, who, ignorant that the final blow had fallen, was anxious to express his sympathy with the affliction of a much-valued friend.

Mon cher Monsieur Stephenson,—Je viens d'apprendre de M. Hugh Thurburn que, depuis quelque temps, votre santé avait été altérée, et je ne veux pas laisser partir ce courrier sans vous dire combien j'ai été affecté de cette fâcheuse nouvelle. J'aime à croire que, loin de considérer ma démarche comme un acte de simple politesse, vous voudrez bien y reconnaître une preuve de l'attachement que je vous porte et de l'estime particulière que j'ai toujours professée pour l'honorabilité de votre caractère et pour votre haute capacité. J'espère aussi que vous serez assez bon pour m'honorer d'un mot de réponse aussitôt que l'amélioration de votre santé vous le permettra.

Agréez, mon cher Monsieur Stephenson, les vœux sincères que je forme pour la conservation de vos jours, qui ne sont pas moins précieux pour notre pays que pour le votre, et recevez l'assurance de mes sentiments les plus affectueux.

MOHAMMED SAID.

Caire, le 17 octobre 1859.

In London and in the North of England the emotion was more personal than elsewhere. In the metropolis, as an eminently popular member of Parliament, as Commissioner on the Health of Towns, a Commissioner for the great Exhibition, the consulting engineer of several railway companies and commercial associations, and an active member of learned societies, he necessarily left behind him a larger number of personal acquaintances than are claimed by most men of active habits. And to have talked with Robert Stephenson once was to feel a cordial affection for him ever afterwards. But even in places which he had never visited, and by men whom he had never seen, he was mourned for, as a great friend snatched away, rather than as a great man summoned to eternity. Strangers had looked upon him in his works, and had surveyed his career again and again in that powerful biography which has made George Stephenson loved by thousands who will never have an opportunity of examining the monuments of his industry.

To the enquiry where Robert Stephenson should be buried, there was only one answer. Without a dissentient voice, pub-

lic opinion demanded the last and highest honour accorded
to great Englishmen—sepulture in Westminster Abbey.

As its cost was not met by the public purse, his funeral
may in one sense be regarded as having been private, but
in all other respects it was a national solemnity. The
authorities having appointed a place for the interment,
near his precursor Telford, application was made by Mr.
Manby, F.R.S., Honorary Secretary of the Institution of
Civil Engineers, to the Duke of Cambridge, to permit the
cortège to pass through Hyde Park on its way to the
Abbey. Without an hour's delay the Duke hastened to
consult the Queen on the subject; and the request was
by Her Majesty made an occasion for expressing her sense
of the loss which her nation had sustained in the following
graceful letter:—

Horse Guards: October 20, 1859.

Sir,—Before sanctioning your proposal that the Funeral
Procession of the late Mr. Stephenson should be allowed to pass
through Hyde Park on its way to Westminster Abbey, His
Royal Highness deemed it expedient to take Her Majesty's
pleasure on so unusual an application, and for which no pre-
cedent exists. Her Majesty considers that as the late Mr.
Stephenson is to be buried in Westminster Abbey, in acknowledgment of the high position he occupied, and the world-wide
reputation he had won for himself as an Engineer, his funeral,
though strictly speaking private, as being conducted by his
friends, partakes of the character of a public ceremony; and
being anxious, moreover, to show that she fully shares with the
public in lamenting the loss which the country has sustained
by his death—she cannot hesitate for a moment in giving her
entire sanction to the course which His Royal Highness the
Ranger recommends.

I have the honour to be, Sir,
Your obedient servant,
(Signed)   J. Macdonald.

Charles Manby, Esq.
&c. &c.

On Friday October 21, the body was conveyed to the tomb. Shortly before 11 o'clock, A.M. the procession, consisting of the hearse drawn by six horses, mourning carriages. drawn by four horses each, and a long line of private carriages, formed in Gloucester Square, and wended its way slowly in the direction of the Abbey.

Entering Hyde Park at the Victoria Gate, the *cortége* moved past the Serpentine to Apsley House, down Grosvenor Place, and along Victoria Street, ultimately pausing in the Cloisters of the Abbey. Throughout the route dense throngs paid, by their quiet demeanour, a genuine tribute of respect to the dead.

The great west door receives the mortal remains only of monarchs and nobles. Borne into the Abbey by a side door, the body of the untitled engineer was received by the dean and clergy, attended by the choir. The pall was borne by the Marquis of Chandos, Mr. G. C. Glyn, M.P., Mr. Joseph Locke, M.P., Sir Roderick Murchison, Mr. Samuel Beale, M.P., and Mr. John Chapman.

On the procession approaching the Abbey, the council and officers of the Institution of Civil Engineers advanced in order from the Jerusalem Chamber, where they had assembled, and took rank with the mourners. In the cloisters also, ready to join in the line, were deputations from the Grand Trunk Canada Railway, the Electric Telegraph Company (of which the deceased was chairman at the time of his death), the Victoria Dock Company, the Royal Society, the Royal Geographical Society, the Geological Society, the Society of Arts, the Meteorological Society, the Astronomical Society, the Birmingham Institution of Mechanical Engineers, and the directors and officers of many railway companies.

So great was the anxiety of the public to obtain admittance to the Abbey, that the approaches were thronged from an early hour by persons seeking entrance. More than two thousand tickets of invitation were issued, and it was computed that at least three thousand persons were present. In that vast assemblage were the best and wisest of the land. Nor was long time to elapse before three of the distinguished crowd followed their friend to the unseen world. Since that day, Joseph Locke, Professor Baden Powell, and Professor Eaton Hodgkinson have been taken from the pursuits by which they conferred enduring benefit on their fellow-men.

On the day of the funeral, the shipping on the river carried their flags half-mast high. At Newcastle, Gateshead, North Shields, Tynemouth, and Sunderland, merchants' offices, banks, and shops were closed at noon, the shipping on the Wear and Tyne paying the deceased the same honour as the fleets upon the Thames. At Whitby the mourning was universal. Other towns showed no less respect to the dead. And at Newcastle, the workmen from the factory of Robert Stephenson and Co., more than fifteen hundred strong, marched to St. Nicholas' Church, and attended divine service.

This strong mark of respect was paid by the men spontaneously, and so much pleasure did their conduct give to Mr. George Robert Stephenson, that two months later he wrote the following letter to Mr. William Pearson, Treasurer of the Workmen's Sick Fund at Messrs. Robert Stephenson & Co.'s Engine Manufactory, Newcastle.

24 Great George Street, London:
Dec. 16, 1859.

DEAR SIR,— I am desirous to express my sense of the admirable way in which the workmen showed their respect and

affection for my late cousin Mr. Robert Stephenson, on the day
of his funeral. When I first heard of their request, that there
should be a special service in St. Nicholas' Church, at the
hour of his interment in Westminster Abbey, and of their
wish to attend in a body, for the purpose of paying the last
tribute to his memory, I felt that such an act, besides being
most gratifying to myself, and the other friends of the deceased,
also reflected the highest credit on the men themselves, and
demanded from me a marked acknowledgment. I therefore
propose to contribute to the Workmen's Sick Fund the sum of
£500, with the understanding that this amount shall be under
the control of trustees, to be invested for the benefit of the
fund. I shall feel obliged by your communicating this to the
men, with my personal thanks for their conduct on the melancholy occasion referred to.

I remain, dear Sir,

Yours faithfully,

GEORGE ROBERT STEPHENSON.

Another expression of regret for the great commander
of workmen deserves record. On the day preceding the
funeral, an urgent application for a card of admission to
the Abbey during the ceremony, was made to Mr. Charles
Manby. The applicant was Henry Weatherburn, a
workman employed on the South-Eastern Railway, who
based his request on the fact that, many years before, he
drove the first locomotive engine, called the 'Harvey
Coombe,' that was used in the construction of the
London and Birmingham Railway. It is needless to say
that the merit of the claim was recognised.

Robert Stephenson's grave in Westminster Abbey occupies a spot nearly in the centre of the nave, near the grave
of Bell, the founder of the Madras system of teaching,
Telford the engineer, and John Hunter the physiologist.

The plate on his coffin-lid bears this inscription—

ROBERT STEPHENSON, ESQ. CIVIL ENGINEER,
D.C.L. AND F.R.S.
Born November 16, 1803.
Died October 12, 1859.

The monumental brass which has recently been placed over the site of his tomb is surrounded by this legend:—

Sacred to the memory of ROBERT STEPHENSON, M.P., D.C.L., F.R.S., &c., late President of the Institution of Civil Engineers, who Died Oct. 12, A.D. 1859, aged 56 years.

A memorial window in honour of the Engineer has recently been placed in the Abbey by his executors, and a bronze statue, by Baron Marochetti, will be erected in the vicinity of Great George Street; that street which, by future generations, will be associated with the name of—ROBERT STEPHENSON.

APPENDIX.

———◦◦———

I. THE CONTROVERSY OF THE GAUGES.

IN ORDER to carry out their undertakings on the Oxford and
Rugby line, in accordance with the privileges preserved to them
by section (5) of the Gauge Act, the Great Western Railway
Company asked their engineer, Brunel, to deliberate on the
subject, and furnish them with a Report thereon. In compliance with this order, Brunel, at the commencement of the
year 1847, sent in the following statement of his views to the
Company.

18 Great Duke Street, Westminster:
January 16, 1847.

GENTLEMEN,—I have deferred reporting to you upon the mode
in which I should recommend to introduce the Narrow Gauge
upon the Oxford and Rugby Railway, because, as the time was
not yet arrived for laying the rails, it was not absolutely necessary; and as whatever mode I might recommend would, as
matter of course, become the subject of criticism by those who
resist the mixture of gauges, as affecting the question of the use
of the Broad Gauge, I did not wish at an earlier period than
necessary to give this new occasion for introducing into any
question that might come under discussion, either before the
public or before Parliament, any part of the old and vexatious
questions of Gauge. As, however, it is desired by the Board of
Trade that some particular plan should now be determined
upon, I beg to say that, *my present intention is to recommend
on the Rugby Line, a single additional rail to each line of
rails, and that the outer rail of each railway should be*

common to the two gauges, in the manner shown in the accompanying plan.

This is the simplest mode of uniting the gauges, and, if it be so desired, admits without any application whatever, of the running of all the trains of both gauges into the same sidings, and up to the same passenger platforms, and, if the switches be coupled or connected, exactly as if no diversity of gauge existed. This will be apparent by the accompanying plan (No. 1), while, at the same time, an advantage may actually be derived from the Double Gauge, which may be made the means of keeping two classes of traffic totally distinct, if such be desired, without the possibility of the trains belonging to the one running into the sidings used by the other when one train has to pass the other, an arrangement which would be particularly applicable to the Rugby Line, if, as has been assumed by the advocates for laying the Narrow Gauge on this line, a large through coal trade is to come upon it from the Northern Gauge Lines.

Thus, as shown on plan No. 2, a siding into which the Narrow Gauge trains would run, and the switch for which could not effect the Broad Gauge trains, which could therefore run past at full speed, without having any switches to attend to; and the Narrow Gauge lines might be carried in like manner past any siding intended solely for the quick Broad Gauge trains, as shown in the same figure.

Crossings from one line to the other are, at least on the Great Western lines, of rare occurrence, and but little used; and on the Great Western Railway they are always kept locked, and used only in case of accident on the stoppage of one line. They are upon all railways placed leading backwards, so that they cannot, even if the switches be wrong, affect the running trains, and are only used at very low speeds. If, therefore, there were any such complication as has been alleged, it would be quite unimportant; but the plans show that they may be perfectly simple, and as easily worked, when they may occasionally be required, as with the Single Gauge. By this arrangement, there is only one switch and point on each main line, but even that, as has been stated, being placed in the direction of, and not meeting, the running train, may be disregarded. The other switch is not on the main line, and moreover, would be self-acting, and being moved by an opposite check rail, would open and shut of itself, according to the gauge of the wheels running past it. Such a crossing, therefore, would be as nearly as possible as simple in respect of each gauge, as an ordinary crossing.

In the plan is shown a crossing, such as we prefer them on the Great Western Railway, combined with the siding or lay-by. As it is seldom that the emergency arises from using a crossing, without at the same time rendering a siding a useful aid in this arrangement, the crossing may be made either common to the two, or the advantage of separating two classes of traffic may be attained, as before, by working the traffic on different gauges, and keeping the crossings as well as the sidings distinct.

Although I have recommended as a general principle, on the Rugby Line, the adoption of a third rail only, the making the outer rail the one common to the two, yet there may be parts of the line where I should prefer two additional rails, making a distinct Narrow Gauge with the wide, as shown in Fig. 6.

With respect to the mode of laying the rails, either longitudinal timbers, as on the Great Western Railway or transverse sleepers, as on the London and Birmingham Railway, may be used. In either case, the mode of laying is not affected by this circumstance of the additional rails or rail. Either three or four longitudinal timbers, each with its rail, may be framed together, just as two are now, and as three or four, or even more now are at crossings or in stations; or three or four chairs may be bolted on one sleeper. It is in this latter mode that I have arranged with Mr. Barlow, the engineer of the Midland Railway Company, to lay the Mixed Gauge, which was originally provided for by Parliament between Cheltenham and Gloucester; but where I should be uninfluenced by the desire to meet the view of others, I should give the preference to longitudinal timbers.

I am, Gentlemen,

Your obedient servant,

I. K. BRUNEL.

To the Directors of the
Great Western Railway Company.

Such were Brunel's opinions on diverse gauges, when the enquiry was shifted from their relative advantages and disadvantages, to the best method of combining. It is to be remarked with what characteristic lightness the undaunted and brilliant engineer passes over difficulties, and brings out what he conceives to be the necessary points of the case.

Stephenson took a very different view of the same subject. At the request of the Secretary for the London and North Western Railway Company, he took pen in hand, and laid before the Directors of that Company and the public, the following criticism on his distinguished rival's Report.

24 Great George Street, Westminster:
July 7, 1847.

DEAR SIR,—In compliance with the request contained in your letter of the 17th April, I have considered the Report of Mr. Brunel to the Directors of the Great Western Railway Company, which has been submitted to the Commissioners of Railways, in explanation of the mode in which it is proposed to carry into effect their order, that the Narrow Gauge (as well as the Broad) shall be laid down on the Oxford and Rugby Railway, and also between Worcester and Wolverhampton, on the Oxford, Worcester, and Wolverhampton Railway.

From this Report, it appears that Mr. Brunel proposes,—

First. Generally to lay the railways in question with one additional rail for the Narrow Gauge, this being the simplest plan; but, wherever it may be more advantageous, to lay two additional rails.

It appears that Mr. Brunel prefers to adhere to the longitudinal sleepers used on the Great Western line, having varied from that system between Cheltenham and Gloucester, rather in deference to the views of others, than in accordance with his own views.

Second. Mr. Brunel proposes, either to mix the sidings and crossings for the two gauges, or to keep them separate. It is not stated which of the plans is preferred; but the feasibility of either or both, where desirable, is affirmed.

It is difficult to reply to this Report in its present form; for though, under the first heading, the main difficulties as regards the laying of the rails are touched upon, they are not solved, and the diagrams do not exhibit clearly the complications which arise from the introduction of the second gauge, being merely rough sketches, and not scale representations of crossings, &c., as actually laid in practice; and under the second heading, no attempt is made to explain the mode proposed for carrying on the traffic at the stations.

I have accordingly prepared more detailed developments of the diagrams * accompanying Mr. Brunel's Report, in order to show the extent to which the rails are actually affected by the intersections which he exhibits by mere cross lines. I have also added plans of approved stations actually executed by Mr. Brunel and myself, composed of various parts, such as the first-mentioned diagrams exhibit separately. Certain other drawings are appended, which will be explained in their proper places. To all these drawings I have preferred to add explanatory notes, rather than to encumber these general pages with detail, which can only be understood by a careful study of the plans themselves. I would remark that, without attention to these notes, it is impossible thoroughly to appreciate the difficulties which are alluded to in these observations.

I now proceed to consider the difficulties attending the introduction of a second gauge, both in the construction and working of a railway, and the best means of removing them, so far as may be practicable.

I trust that the length to which my remarks will necessarily extend, will appear to be justified by the importance of the subject. You will remember not only that the Great Western Railway Company are bound, by the requirements of a Parliamentary Committee, and of the Board of Trade, to maintain a Narrow Gauge Railway, with all facilities for working it, at the stations, and along the line from Rugby to Oxford; but they have further pledged themselves to connect Oxford with the Narrow Gauge South Western chain of railways at Basingstoke, by means of a Narrow Gauge link through Reading.

The establishment of an unbroken connection, by means of a

---

* It has not been thought requisite on the present occasion to reproduce the diagrams of the two engineers. For general purposes the arguments of either writer can be followed with sufficient accuracy without the illustrations. The professional reader and the curious can consult the full Report, published in 1847.

T

safe and properly-constructed Narrow Gauge Railway, between
the two Narrow Gauge systems of railway, in the North and
South of England, is therefore considered by the legislative
authorities a matter of national concern, for otherwise the
construction of a duplicate railway 90 miles long would not be
enforced.   During the present session, moreover, not only have
about 150 additional miles of railway been proposed to be laid
on the Mixed Gauge system, but various Parliamentary Committees have appeared to consider it important to insert in
Railway Bills a clause, binding Companies to lay down
additional rails for another Gauge, on the Railway Commissioners deciding to exercise powers to order such additional
rails.

Since, then, the notion of this untried system of Mixed
Gauge, and the reliance upon it as a remedy for the evil of a
Break of Gauge, and as a compromise between the conflicting
systems of Gauge, seems to be spreading both in Parliament and with the public, I am desirous not to let pass the
present opportunity of treating the subject as completely as
our present experience permits.

Permanent Way.

In considering the effect of introducing a mixed gauge as
regards the construction of the permanent way, I omit, for the
sake of brevity, all remarks on the mode of laying the
additional rail along the main line where it is unbroken by the
crossings, sidings, &c., necessary for the carrying on of traffic at
stations and other points.   The mechanical possibility of laying
down an additional or intermediate rail in such parts of the line
is admitted, and the expense of laying it down is given in
Appendix II.; I am, however, of opinion, that the keeping the
road in repair under the three-rail system, even in those parts
of the line where there are no intersections of the rails, will be
troublesome and expensive, in consequence of the unequal action
upon the rails.

Premising this, I proceed to consider the difficulties attending those points of divergence from, and intersection with, the
main line, which are necessary accompaniments of every railway
serviceable for traffic.

First, then, following the order of Mr. Brunel's remarks, as to crossings. My whole experience of railway arrangements leads me to a conclusion very different from Mr. Brunel's, that crossings between the main lines are but rarely necessary. On the London and Birmingham Railway, for instance, a length of 112 miles, there are no less than 58 crossings, and on the lines on the north, where the mineral traffic is greater, their number in proportion to the length of the railway is much increased. Indeed, wherever we engage to admit traffic on a railway, whether at passenger stations or at intermediate points along the line, such a crossing between one main line and another is necessary; the ordinary process of returning a carriage or wagon from any station cannot be effected without the intervention of such a crossing, and if a station is only on one side of a railway, as is sometimes necessarily the case, say the *up* side, the traffic from such a station may pass *up*, without crossing the *down* line, but cannot pass *down* without crossing the up line; so that on this account also a crossing is necessary. In one way or other, then, crossings are necessary wherever traffic arises; from the annexed plan of the Slough station (drawing No. 7), laid out by Mr. Brunel, it will be seen that at this station only there are two crossings between the main lines.

With reference to the mode of laying a through crossing on the three-rail system, I find, after the most careful attempt to introduce into the crossing the automaton switch (in the manner proposed by Mr. Brunel for uniting the gauges), that room cannot be found for its efficient working (see drawings 4 and 4A); it will therefore be necessary to use additional half switches (at A and B, Fig. I., drawing 4), for this purpose, rendering two switch-men necessary to work a crossing instead of one, and introducing into the Narrow Gauge traffic new chances of accident, arising from any failure of the two men working together.

In stations laid out as at Slough and Reading, on one side of the line, so that every train entering the station has to meet the switches, the risks attending these complexities of arrangement would be most serious.

On the same drawing (No. 4), the alternative of separate crossings for the two gauges is suggested, an arrangement which, from its simplicity, is preferable to that of mixed crossings, but requires more space.

It is to be observed that the switches are by no means the only objectionable parts of a crossing, a junction between two lines of railway, a siding, or any other point where one system of rails necessarily intersects the other. Wherever another rail crosses the rail of the main line, a *gap* necessarily occurs in the main line, in order to allow the flange of a wheel to pass as it goes along the cross line; every such gap inflicts a blow on every wheel that passes over it, and the higher the speed the greater the blow. How much of the wear and tear of stock arises from this source, it is perhaps impossible to ascertain with precision; but it is notorious that the wear and tear of a most important part of the machinery of a railway is mainly due to it—that of the wheels and axle. An injury thus sustained may not cause an accident as the wheel passes over the spot where the blow is inflicted, but may yet lead to one at another place, the cause of which cannot then be determined. The bending, or partial fracture of an axle, sets the wheel out of gauge, and prepares for greater straining at the next point where the main line is intersected; and thus, by a succession of blows, breakage ensues, or, even without breakage, the carriage is thrown off the line.

Again, the liability to run off is greatest at these places of intersection, or gaps in the main rails.

From the subjoined record (Appendix I. A) of the accidents which have occurred on the Midland Railway during the last twelve months, kept by Mr. Barlow, the resident engineer, it appears that on that line seven-eighths of the whole number of accidents of a most dangerous class (namely, those resulting from trains getting off the rails) have arisen at places where the continuity of the rail has been interrupted by switches or crossing points. To increase the number of such gaps is, therefore, to add proportionably to what experience has shown to be the most frequent cause of accidents of the most serious class.

In Appendix I., the results of my examination of Mr. Brunel's diagrams are collected in a tabular form, showing the increased number of such gaps in the continuity of the main lines of rail, and of fixed points to be met by the trains, in stations, at junctions, and at sidings, by the introduction of one additional rail, and by that of two additional rails.

The description of switch referred to in this table, is the single switch, shown in Fig. 1, drawing A.

This description of switch has been in most cases discarded, in consequence of the disadvantages pointed out in Fig. 3, drawing A. It is assumed in these remarks, as the switch adopted for the Mixed Gauge, because it is the form suggested by Mr. Brunel as the most applicable.

With regard to the advantages claimed by Mr. Brunel as attending the use of this switch under certain circumstances, it is only necessary to state, that these advantages, such as they are, are purchased by the adoption of an inferior form of switch, involving an additional gap and fixed point in the common main rails. (See Fig. 3, drawing A.)

Referring to the table of particulars (Appendix I.), it is only necessary here to state generally, that

In the case suggested in drawing No. 1 (of Mr. Brunel):—

There will be two additional half switches, two additional crossing points, two additional pairs of over-crossing points, four additional gaps, and three additional meeting points, to be passed over by the trains of either one gauge or the other, or by both.

In the case of drawing No. 2 (of Mr. Brunel):—

There will be switches at four places instead of two, as with the Single Gauge, and there will be two additional crossing points, two additional pairs of over-crossing points, six additional gaps, and four additional meeting points, to be passed over by trains of either one gauge or the other, or by both.

In the case of drawing No. 4 (of Mr. Brunel):—

There will be two automaton switches of a novel and unsafe construction (see drawing 4, A), to be passed over by every train using the crossings, one of these switches being placed the wrong way, so as to meet all the trains in one direction. There will also be two additional half switches, four additional crossing points, two additional pairs of over-crossing points, six additional gaps, and four additional meeting points, to be passed over by trains of either the one gauge or the other, or by both.

In the case of drawing No. 5 (of Mr. Brunel):—

There will be two automaton switches, as in drawing No. 4, to be passed over by every train using the crossing, one of these switches being placed so as to meet all the trains in one direction; there will also be five additional crossing points, four additional pairs of over-crossing points, nine additional gaps and four additional meeting points.

In the case of drawing No. 6 (of Mr. Brunel):—

There will be switches at four different places instead of at two, as with the Single Gauge; there will also be six additional crossing points, four additional pairs of over-crossing points, twelve additional gaps, and eight additional meeting points, to be passed over by trains of either one gauge or the other, or by both.

All these additional points of danger, arising from additional intersections of the rails (with the exception of the automaton switches in drawings 4 and 5), occur on the main lines of rails; so that an accident arising at one of these points would affect the traffic elsewhere, and delay and embarrass, in all probability, other trains than that to which the accident occurred.

The three-rail system thus increases the interruptions or gaps in the main lines of rails, and the still more objectionable meeting points, in the proportion of about two to one, and the four-rail system about as three to one. But if we compare the results of the record of accidents on the Midland line with the table of additional intersections, and take into account the necessary increase in the number of trains, attending the use of a Mixed Gauge, referred to hereafter in these remarks, under the head 'Traffic Arrangements,' I consider that the three-rail system would increase the danger in the proportion of at least four to one, and the four-rail system as five to one, from that cause which has been proved by experience to produce seven-eighths of the railway accidents consequent upon engines or carriages running off the line.

It is further considered, that the more closely these interruptions of the main line occur, the more difficult it will be to keep the line in good working order, and this far beyond the direct proportion of the number of such interruptions.

From drawing No. 9 (showing a junction laid with two rails, with three and with four) it will appear that, as the number of crossing points increase, the main line rail becomes cut up into shorter lengths, and the shorter the length of rail, whether of Mr. Brunel's bridge rail, or longitudinal timbers, or of the double Trail on cross sleepers, the more difficult it necessarily becomes to fix and maintain it.

From the same drawing (No. 9), and from the other drawings,

it also appears that the check rails, at present used as palliatives of danger, become more difficult to apply, just as the danger or the necessity for such check rails increases; there being in some cases no room at all for them, as, for instance, in such complications as arise in some portions of the Slough station, and at over-crossing points generally.

In the Mixed Gauge system, therefore, we are partially deprived of a palliative, which is deemed essential to safety under the Single Gauge system, although we need it more than under the ordinary system.

The danger encountered by a fast train running through a station without stopping, will, from these causes, be greatly aggravated.

From the whole of the preceding considerations, the following general conclusions, in reference to the effect of the Mixed Gauge system on the permanent, are drawn.

First. That the four-rail system increases the complication so much, as to be inadmissible, except where absolutely necessary, as at turn-plates, &c.

Second. That notwithstanding the exercise of the utmost care, in construction and maintenance of way, the mixture of gauges, either by means of three or four rails, introduces in the road itself a greatly-increased risk of accident, entirely incapable of remedy, and scarcely justifiable by any considerations of mere convenience.

Arrangements of Stations.

In considering in the next place the arrangements of stations, which will be necessary on a railway of Mixed Gauges, the question of mixed or separate sidings and crossings must be considered more fully than is done in Mr. Brunel's Report.

It may be stated at the outset, that the separate system will require separate switches at separate places, instead of the connected half switches added by the mixed system, while it will also introduce an increased number of crossing points; but the separate plan will render them more simple, and more easy to lay with proper checks, and to keep in repair.

The idea of arranging Broad and Narrow Gauge stock promiscuously together, suggests immediately great inconvenience.

As regards spare stock, the separate system of arrangement is,

on the face of it, the more convenient, because, whenever an
extra vehicle is wanted, its gauge being an essential part of the
demand, less shifting of the stock will necessarily be required,
if the two classes are kept separate, than if they are placed
promiscuously together; labour and time will thus be saved by
separating them.

Common sidings for both gauges could only be used alter-
nately for vehicles of the one gauge or the other, not for both
at one time, but if one such siding were laid down, there could
be no security against their being used for both kinds of stock
at the same time; it is therefore necessary clearly to point out
the danger which would arise from such a practice.

This danger will be readily appreciated by any one who has
stood by and observed a carriage at a station come in contact
with another, while a porter stands perfectly secure between the
buffers, the carriage bodies being prevented, by the protection
of the buffers, from approaching so close as to crush him.

By reference to drawing (No. 10), it will be seen that the
carriages of the ordinary construction, Narrow and Broad
Gauge, when standing together on a common three-rail siding,
will not buff together, the buffers missing each other, and
coming in contact instead of with the opposite carriage frames.
It is unreasonable to expect that a porter shall always, in the
hurry of business, be able to consider whether an approaching
carriage is Narrow or Broad Gauge, and, consequently, whether
he is safe or not if engaged in the siding; at any rate, no such
danger should be thrust upon him by defective arrangements.
To render a three-rail common siding safe, it would then be
necessary to alter the buffers of the entire stock of the whole
connected system of railways, using such sidings in such way,
that the narrow and broad carriages should buff together; for
when once a number of railways are connected in one system,
the stock finds its way to every part of the system at one time
or other; indeed, it is the very object of the proposed mixture
of gauges, that goods and passengers may proceed to their
destination without change of vehicles; Narrow or Broad
Gauge vehicles must be everywhere in turn, as the traffic
demands; and, until all of them are made to buff indiscrimi-
nately together, there can be no safety for porters in a three-rail
siding. Accidents of this nature occurred from the use of stock

not buffing together, on the Glasgow and Greenock Railway, until the remedy was applied; within a short period, several persons were injured, some fatally.

Besides the danger to the men, it is clear that the buffers of every carriage (whenever two of different gauges come together), striking the cross framing of the other carriage, instead of the longitudinal framing, or buffer ends, would knock the carriages to pieces; one or two blows of this nature, indeed, would in all probability so shake the framing of the carriage as to make it unsafe.

Again, in the moving about of stock at stations of considerable traffic, engine power is used with much advantage. In the same drawing (No. 10), a Narrow and Broad carriage are shown coupled, to draw attention to the fact, that in a three-rail siding used promiscuously for both kinds of stock, this economical and necessary power cannot be used, without much danger of throwing the vehicles off the rails by the oblique line of traction which results.

Whenever turn plates are required, the three-rail system must be converted into the four-rail system, as the action of a turn-plate implies a concentric arrangement.

For these reasons, the application of the three-rail system at stations appears inadmissible.

It may here be remarked, that collisions on a three-rail Mixed Gauge line, between trains of different gauges, would necessarily become more disastrous in their consequences than under the present system; the blow would be oblique, and would, therefore, tend to twist and double up the train, and so to bring the carriages into a position in which they would be least able to resist the crushing force.

The four-rail system for sidings has already been shown to involve a greatly-increased number of dangerous points of intersection. It would also be inconvenient at the platforms.

The only alternative, then, is to lay separate sidings for the two gauges, or, in other words, completely to duplicate the station arrangements.

Traffic Arrangements.

I now proceed to consider how far the number of trains required is affected by the mixture of gauges.

The object of a Mixed Gauge professes to be, to afford to the public the same amount of convenience as they would derive from an uniform system of one gauge. Under such a system, passengers or goods from Banbury (for instance), either northward to the Narrow Gauge district, or south-westward to Bristol, or other places now on Broad Gauge lines, would be able to proceed, without change of vehicle, by any train of the day which might be most convenient. To retain the same amount of accommodation on a Mixed Gauge line, it is clear that the same number of trains daily would be required of each gauge, that is to say, the amount of trains collectively, would need to be doubled. The stock would therefore be nearly doubled, the amount of labour and expense in despatching the trains would be much increased, the risk of collisions would be fully doubled, and the profit of course would be much diminished, as the public accommodation would not be (by the supposition) increased, but would only remain the same as it would be under an uniform gauge.

The interest of a Company possessing a Mixed Gauge line would therefore clearly be, to reduce expense by working mainly on one gauge, choosing whichever might be most in accordance with their general interests; in which case, it is clear, that they would no longer afford the public (as was implied by the proposal of a Mixed Gauge line) the same advantages which would have been derived from a separate line of that other gauge, the working of which would inevitably be, for the reasons we have considered, discouraged.

Cost.

Approximate estimates of the expense arising in various ways from the mixture of the gauge, are appended. By Appendix II. it will be seen that the additional expense arising from laying the Narrow Gauge (with the necessary facilities for working it) on a Broad Gauge railway, would be about £5,974 per mile, while, by Appendix II., the increased annual expense of maintenance and working, exclusive of interest on additional capital, would be at least £500 per mile per annum. This latter estimate is founded both on my general experience of the expense of working and maintaining railways, and on a return of the actual cost per train per mile on the London and Birmingham line, where, from the great number of trains, the cost *per train* is low.

If, as is generally done in such cases, we capitalise this annual expenditure of £500, it will amount (at 4 per cent.) to £12,500; adding this to £5,974, the total additional cost of a Mixed Gauge line would be £18,474 per mile.

For this vastly additional outlay, we should get an inferior railway, less safe, and less efficient than the ordinary form of railway.

Looking at the question of cost in another light, and having regard to the fact that one principal reason of the establishment of the Narrow Gauge on the Oxford and Rugby line, was the accommodation of the coal traffic of the north; this extra annual expense of £500 per mile for working would, if saved, be sufficient for the carriage of 120,000 tons of coal per annum over the whole length of the railway free of all charge, or adding 4 per cent. interest on the original outlay of £5,974, for the carriage of 177,360 tons.*

Or, applying the result to passengers, the whole average number of railway passengers travelling per mile per annum, is about 180,000; a Railway Company might therefore take off 1*d.* per mile from all its passengers' fares, for the same annual sum as the laying and working the additional gauge would amount to.

I have in the foregoing remarks endeavoured fairly to appreciate, and, as far as possible, to point out the means of obviating, by the separation of the stations and traffic, the difficulties both of original construction, of maintenance, and of working, arising from the mixture of gauges, as well as to estimate the probable additional expense of the system.

What, then, are the advantages to result from its adoption?

By a Narrow Gauge line from Rugby, through Oxford to Reading and Basingstoke, the through traffic from the Narrow Gauge districts in the north and north-east, will be carried without interruption to the south, south-west, and south-east, and the communication of the agricultural counties, through which the Oxford and Rugby line passes, with the coal-fields of Leicestershire will be effected.

---

* Over the whole Railway, allowing one penny per ton per mile for the expense of carriage.

On the other hand, the only ostensible object of the Broad
Gauge on the line between Oxford and Rugby, is the connection
of Exeter, Bristol, and the agricultural districts passed through
by the Great Western Railway, with the towns of Banbury and
Rugby, and with the district (also agricultural) between Oxford
and Rugby, and this local object, we have seen, can only be
purchased at a great cost, and by a great sacrifice of safety.

In the case of the Oxford, Worcester, and Wolverhampton
line, a main design of which is to afford accommodation to the
owners of mineral property in conveying their heavy produce to
the works of their customers, the increased cost of the Mixed
Gauge arrangements will to a great extent, if not entirely, defeat
the local object which it is proposed to effect.

Stock for both gauges will have to be furnished and maintained; extra lines for the connection of the works with the
railway to be laid down, and extra space for the carrying on of the
traffic to be provided; this, where the cost of the carriage forms
a large part of the price of the article when delivered to the
consumer, will clearly be a heavy tax on the producer, if it does
not, indeed, in many cases, amount to an absolute prohibition
of railway conveyance.

It is important to observe, that the termini of this Mixed
Gauge system, introduced to remedy the evils of the Break of
Gauge, become themselves again the sites of other breaks of
gauge, at one or other of which all through-traffic from a
Narrow Gauge to a Broad Gauge district, or *vice versâ*, must
be transferred. Thus, by the Mixed Gauge system, the evils
which have hitherto been confined to one district will be
repeated in new and distant parts of the country.

The Break of Gauge, in short, cannot be remedied by the
mixture of gauges, until such mixture becomes co-extensive
with railways themselves, and thus, after enormous expense, we
should arrive at last at uniformity, but with the simplicity of
construction and arrangement, which can alone ensure the
economy, safety, and efficiency of the railway system.

I am, dear Sir,

Yours truly,

ROBERT STEPHENSON.

To R. Creed, Esq.
Secretary to the London and North Western Railway Company.

This masterly Report (which, with all its strength, complete-
ness and accuracy, may be regarded as nothing more than a
fair sample of Robert Stephenson's professional and official
papers) swept away the last hopes of the Broad Gauge adherents,
cutting from under them the ground on which with characteristic tenacity of purpose, they were rallying their forces for
another fight—and forcing the most prejudiced and most
obstinate opponents of the Narrow Gauge to admit that their
cause was lost beyond a hope of recovery.

II. INTRODUCTORY OBSERVATIONS ON THE HISTORY OF THE BRITANNIA AND CONWAY BRIDGES.

By Mr. ROBERT STEPHENSON.

SHORTLY after the metropolis of this country and the great
commercial port of Liverpool had been connected by means of
railways, the public attention began to be directed to the further
improvement of the communications with Ireland. This, it was
obvious, could only be accomplished by extending the land
journey and diminishing the sea voyage; or, in other words, by
increasing the comparatively certain and diminishing the
uncertain portion of the journey. The ports of Holyhead and
Dynllaen each had their advocates, as the most eligible packet
station and terminus for a railway destined to curtail the journey
from London to Dublin. The relative merits of these two points
of departure for Dublin were keenly discussed, and various investigations were entered upon, and reports made, both by civil
engineers and naval officers.

Those discussions terminated in the preponderance of evidence being in favour of Holyhead, which led to the adoption
of the line of railway as then designed by my lamented father
the late George Stephenson, which, with the exception of about
five miles in the neighbourhood of Bangor, is that which has now
been brought so near to completion. The first survey for this was
made as early as 1840, but the formal application to Parliament
did not take place until the sessions of 1843-4. The chief
engineering work then involved was the bridge over the River
Conway, close to the existing suspension bridge. The passage

over the Menai Straits was proposed to be effected by permanently appropriating one of the two roadways of the great suspension bridge to railway purposes. The steep ascents at each end of the present suspension bridge it was designed to avoid by elevating the level of the railway to that of the suspended roadway at its highest point. As the strength of the suspension bridge was deemed inadequate for carrying safely railway trains and ponderous locomotive engines, it was intended to convey the trains across in a divided state, if necessary, by means of horse-power, another locomotive being in readiness to be attached on the opposite side; thus the passage of engines was entirely obviated. To this proposal the Commissioners of Woods and Forests assented, with the condition, however, that the appropriation of the south suspension roadway for railway purposes should only be temporary. Such a stipulation rendered their assent merely nominal, because the expenses which must necessarily have been incurred in carrying out the proposal were, although suggested with the view of limiting the cost of crossing the Straits, totally inconsistent with the idea of its being only a temporary expedient.

The company were thus driven to abandon this part of their plan and to propose an independent bridge for the railway.

The bill was permitted to pass Parliament with an hiatus of five miles, which were effected by the abandonment of the suspension bridge as a means of crossing the Menai Straits. There were, however, other objections distinct from that just alluded to in reference to the suspension bridge. Some parties urged that this portion of the line was needlessly circuitous—that three-fourths of a mile might be saved by some additional expense. Others objected to it, because it approached and interfered with the privacy of the residence of the prelate at Bangor. It would be out of place here to discuss the value of either of these objections: it is sufficient to say that they prevailed, and the company directed their engineer to deviate the line to avoid them, and to select the best point for crossing the Straits by an independent bridge.

Previous to the erection of the suspension bridge by Telford, in 1826, various modes and points of crossing had been proposed by Rennie and Telford. Their reports, plans, and opinions, were carefully studied, which led to the adoption of the site known by the name of the Britannia Rock, about a mile to the south of Telford's suspension bridge. This spot is peculiarly eligible for the purpose, the rock being nearly in the centre of the channel, rising just to high-water mark, and of sufficient area to admit of the easy erection of a pier upon it. The channel is here unbroken during the ebb and flow of the tide. These peculiarly favourable circumstances were considered highly advantageous, not only for facilitating the erection of a bridge, but for rendering such a structure unobjectionable to the navigation of the Straits. *It was proposed to construct the bridge of two cast-iron arches, each 350 feet span, with a versed sine of 50 feet, the roadway being 105 feet above the level of high water at Spring-tides.*

The span here proposed was the same as that which had from the first been designed for crossing the Conway River.

Such was the state of the engineering problem in reference to the Conway and Britannia Bridges when the company obtained the first Act of Parliament in July 1844. It was proposed to construct a bridge consisting of one arch of the unusual span of 350 feet over the Conway River, at 20 feet above high-water mark, and another over the Menai Straits at the Britannia Rock, consisting of two arches each of a similar span, but at the elevation of 105 feet above high water Spring-tides.

The rise of the tide in both cases is nearly the same; the channels are also very similar, being from 50 to 60 feet deep, with a rocky bottom, and a rush of tide reaching five miles an hour at Conway, and seven miles an hour in the Straits.

These conditions, together with the necessity of keeping the channels open at all times for the purposes of navigation, rendered it perfectly clear that none of the methods heretofore adopted in the erection of cast-iron arches could be brought to bear in either of these localities. The inordinate cost of centering, even if other arrangements had admitted of its application, was at once fatal to its adoption; and it soon became evident that some means external to the arch should be employed to suspend the voussoirs, or ribs, until the arch was keyed in.

A contrivance of this kind had at one time been considered by Telford for the suspension of centering, upon which he proposed to frame and connect the voussoirs, or ribs, of a cast-iron arch; and a slight drawing of such a project is given in the account of the Menai Bridge. Without going into the merits of this proposal in the form suggested, or into its applicability to the present case, it is sufficient to say that it was discarded, and a modification, as brought forward some years ago by Sir Isambard Brunel, for constructing brick arches without centering, taken up as more suitable. Sir Isambard's idea, which was experimentally carried out to a great extent, appeared unexceptionable, and led to the following design for the erection of the cast-iron arches at the Britannia Rock. Instead of two arches being erected upon two abutments and one pier, it was proposed to treat the abutments as piers also, and to complete the iron-work in the form shown by the following figure:—

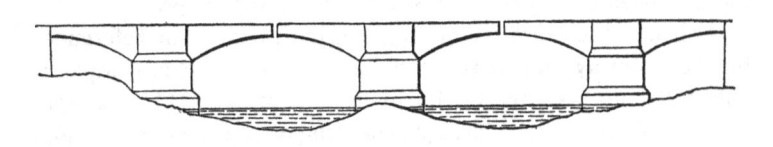

The erection of the arch was to be proceeded with by placing equal and corresponding voussoirs on opposite sides of the pier, at the same time tying them together by horizontal tie-bolts, as shown below.

This system, it is confidently believed, may be successfully carried out to a far greater extent than would have been required in the case of the Britannia Bridge.

It will appear evident, on a little reflection, that as every succeeding step of voussoirs is secured by the tie-bolts, the

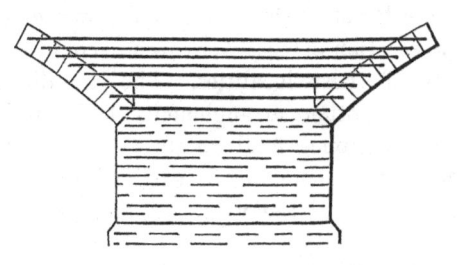

tension of the last bolt, as well as all the previous ones, will be relieved by an amount equal to the whole of the horizontal thrust due from the voussoirs last placed.

If the voussoirs could be constructed or weighted, so that an arch equilibrium could be formed, all the horizontal tie-bolts might be removed, except the last one, for in such an arch the horizontal thrust is everywhere equal. It is not meant that such a method of proceeding as that of removing all the bolts could be carried out practically—it is merely alluded to here to show how largely the bolts would have been relieved from strain as the arch progressed into a form which might appear to endanger the stability of the structure.

Had this plan been carried out, it was not intended to have keyed the arches at the crown, but to have left ample space between the culminating voussoirs to admit of expansion and contraction taking place freely. The bridge would, therefore, have been simply a double-jibbed crane, perfectly balanced on each pier. A connection at the apex of each arch would be necessary, but so contrived as not to interfere in the least with the expansion and contraction, and yet to counteract any tendency to tilt, consequent on the variable pressure of the passing loads.

This mode of construction, although decided upon for the Britannia Bridge, was found unsuited for that of Conway. There only one span was required, and the springing of the arch would have been below the high-water line, and from a natural mass of rock on both sides, which at the east extremity rose nearly to the permanent level of the railway.

It was, consequently, impossible conveniently to treat the abutments in the light of piers, as has been just described. Moreover, the great additional expense of this method, where one arch only is required, formed a serious objection to it, as it necessarily involved the use of double the weight of material requisite for one simple arch, the weight of each overhanging wing being equal to half the weight of the arch itself.

The objection on the score of expense did not apply to the Britannia, for there the overhanging wings were a useful portion of the bridge, and formed a substitute of the extension of masonry, which would have been nearly as costly. Both the expense, therefore, and the peculiarity of the site of the Conway Bridge, pointed out the necessity of some other method being devised for the erection of the arch. Various modes for erecting and supporting a fixed centering were considered, but none appeared satisfactory or safe ; whilst the formidable difficulty of stopping

the navigation, and seriously interfering with many vested
interests for probably two years, remained in all its force.

This state of things led to the idea of building the arch complete on centering supported entirely upon, and framed into,
a series of pontoons kept afloat during the whole time of construction. This arrangement, which is shown in the following
sketch, appeared upon the whole far the most feasible that had
been suggested, and well adapted for placing the arch into its
permanent position :—

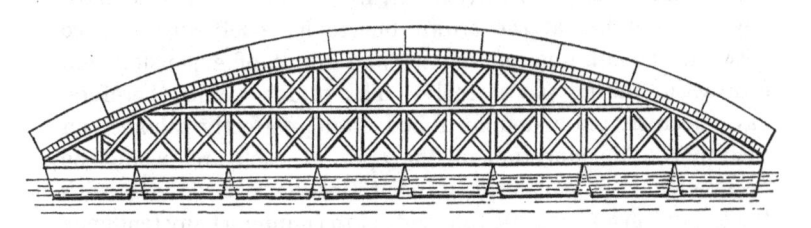

The rise and fall of the tide were such as to admit of its being
brought immediately above the springings and lowered into its
place by the falling tide, or by admitting water into the pontoons
at the top of the tide, before the velocity of the ebb-stream had
increased so as to interfere with the accurate adjustment of the
descending mass. This method of fixing arches I have since
learned was proposed many years ago by Mr. Dixon, of Darlington. He made designs for a cast-iron bridge across the River Tees,
at Stockon, and, instead of erecting centres on the permanent
site of the arch, he proposed to use pontoons, precisely in the
manner which has been described. These plans were not carried
out, in consequence of the Stockton and Darlington Railway
Company having determined to try a suspension bridge for railway
purposes instead of a cast-iron arch. For a brief description
of the particulars of the novel proposal of Mr. Dixon, I have
been favoured with a communication from Mr. R. B. Dockray,*

---

* Euston Station : May 25, 1849.

My dear Sir,—In accordance with
your wish, I beg to send you the
following account of a mode of
erecting the ribs of a cast-iron arch
proposed by the late Mr. James
Dixon, of the Stockton and Darling-.
ton Railway.

The proposed bridge was for the
purpose of carrying the Middles-
borough branch of the Stockton and
Darlington Railway across the River

who resided at Darlington at the time when Mr. Dixon made the design. I have also learnt from Sir John Rennie that this was the method adopted for placing the centering of the Waterloo and London Bridges; the centres being constructed on pontoons, and floated and lowered into their proper position.

Such were my intentions regarding these two bridges when the general meeting of the Chester and Holyhead Railway Company took place on August 30, 1844.

In the following November, the Company deposited new plans preparatory to an application to Parliament in the ensuing session for the deviation which had been forced upon them by the circumstances already alluded to—extending from near the River Ogwen to Llanfair, in the island of Anglesey, comprising, in this distance, a series of railway works unparalleled in cost and magnitude, the Britannia Bridge being one of them.

Tees, at a little higher up the river than Stockton. Several plans were laid before the directors, and at length that of Captain Brown, for a suspension bridge, was adopted and carried into execution. This bridge, as you will remember, proved insufficient for the weight passing over it, and for several years it was supported by timber gearing, carried upon piles driven into the bed of the river, until at length it was' entirely removed, and the present cast-iron-girder bridge, built under your directions, was substituted. One of the designs originally submitted to the directors was by their resident engineer, Mr. James Dixon: the bridge was of cast iron, of three openings of about (if I remember right) 80 feet each. There was nothing particular in the construction of the bridge itself; but it was supposed that the Tees Navigation Company, then hostile to the Railway Company, would object to the temporary obstruction of the navigation by the erection of the centres for

placing the ribs of the arches. To obviate this difficulty, Mr. Dixon proposed to erect each rib separately upon a scaffolding or centre placed in a pontoon, at such an elevation that when floated to the site of the bridge the ends of the ribs would clear the skew-backs on the pier and abutment, and, when properly moored in this position, the pontoon was to be lowered (until the rib rested on the skew-backs) by admitting the water into the hold; and thus he proposed to proceed with the erection of each individual rib.

I saw the drawings in the year 1831 or 1832, not only of the bridge, but of the pontoon, with its centre and valve for admitting the water. They were in detail, and beautifully executed, and I have no doubt they are still in existence, probably in the hands of his brother, John or Edward.

I am, my dear Sir,
Very truly yours,
ROBERT B. DOCKRAY.
To Robert Stephenson, Esq. M.P.

Immediately on its becoming known what description of bridge it was intended to throw over the Menai Straits, a new series of objections was raised, and a violent opposition started on behalf of those interested in the navigation of the Straits. It was urged, that any such bridge as that proposed would seriously injure and fatally aggravate all the evils and dangers which beset the navigation. It was maintained that the difficulties of navigation arising from the great velocity of the tidal currents, the rapid eddies, the number of sunken rocks, and the baffling winds which frequently prevail, demanded the utmost skill on the part of the pilots to avoid accidents of a serious character: hence the necessity of Parliament refusing to sanction the erection of arches, which,.in consequence of the great area occupied by the spandrils and piers, would not only restrict vessels to a narrower channel than heretofore on passing near the Britannia Rock, but would also shelter the vessels from the wind in situations where it was of the utmost importance to them.

These objections were deemed by many, deeply interested in the Holyhead Railway, so grave as likely to endanger the success of the Bill then before Parliament, and consequently the whole undertaking. In this position of affairs I felt the necessity of reconsidering the question, whether it was not possible to stiffen the platform of a suspension bridge so effectually as to make it available for the passage of railway trains at high velocities.

In an attempt of this kind one remarkable failure had taken place some years before at Stockton-upon-Tees, and a professional survey of that structure had sufficiently demonstrated the extreme difficulty of such a task.

At this time Mr. Rendel called my attention to the mode of trussing which he had arranged for preventing oscillation in the platform of suspension bridges, and afforded me the opportunity of inspecting the working drawings of the method he had pursued in correcting that defect in the Montrose Suspension Bridge, which gave way from the accumulation of a mass of people during a boat race, on March 19, 1830, and again, subsequently, during a hurricane, on October 11, 1838. As this latter accident appeared to arise from, or at least to be materially aggravated by, the flexibility of the roadway, Mr.

Rendel, being appointed to repair it, devised an excellent system of trussing, which has stood the test of several years. An elaborate and interesting description of the repairs executed under Mr. Rendel's directions is recorded in the 'Transactions of the Institute of Civil Engineers for 1841.'

The system of trussing here adopted by Mr. Rendel appears to me admirably adapted for a suspension bridge intended for such weights as pass along turnpike roads, but the case under consideration was unquestionably very different, and certainly demanded, if trussing were resorted to, a much stronger and more ponderous system than that followed on the Montrose Bridge. Amongst a variety of devices for the accomplishment of this object, the most feasible appeared to be the combination of the suspension chain with deep trellis trussing, forming vertical sides, traversed by suspension rods from the chains, with cross bracing frames top and bottom, thus forming a roadway surrounded on all sides by strongly trussed framework.

A structure of this kind would no doubt be exceedingly stiff vertically, and has, indeed, been applied successfully in America on a large canal aqueduct, and is clearly described in the 'Mechanics' Magazine,'* vol. xliv. 1846.

---

* This work consists of seven spans, of 160 feet each, from centre to centre of pier. The trunk is of wood, and 1,140 feet long, 14 feet wide at bottom, 16½ feet on top, the sides 8½ feet deep. These, as well as the bottom, are composed of a *double* course of 2½-inch white pine plank, laid diagonally—the two courses crossing each other at right angles, so as to form a solid lattice-work of great strength and stiffness, sufficient to bear its own weight and to resist the effects of the most violent storms. The bottom of the trunk rests upon transverse beams arranged in pairs 4 feet apart; between these, the posts which support the sides of the trunk are let in with dovetailed tenons, secured by bolts. The outside posts, which support the side-walk and tow-path, incline outwards, and are connected with the beams in a similar manner. Each trunk-post is held by two braces; 2½ × 10-in. and connected with the outside posts by a double joist of 2½ × 10. The trunk posts are 7 inches square on top, and 7 × 14 at the heel; the transverse beams are 27 feet long, and 16 × 6 inches; the space between two adjoining is 4 inches. It will be observed that all parts of the framing are double, with the exception of the posts, so as to admit the suspension rods. Each pair of beams is supported on each side of the trunk by a double suspension rod of 1⅛th-inch round iron, bent in the shape of a stirrup, and mounted on a small cast-iron saddle, which rests on the cable. These

The application, however, of this system to an aqueduct is perhaps one of the most favourable possible; for there the weight is constant and uniformly distributed, and all the strains consequently fixed both in amount and direction, two important conditions in wooden trussing constructed of numerous parts. In a large railway bridge it is evident, so far from the conditions obtaining under any circumstances, they are ever varying to a very large extent, but when connected with a chain which tends to alter its curvature by every variation in the position of any superincumbent weight, the direction and amount of the complicated strains throughout the trussing become incalculable as far as all practical purposes are concerned.

Putting these objections on one side for a moment, the introduction of wood into such works as the Conway and Britannia Bridges seemed inadmissible, both on account of its perishable nature and danger from fire.

This led to the revival of a design I had made in 1841 for a

---

saddles are connected on the top of the cables by links, which diminish in size from the pier towards the centre. The sides of the trunk rest solid against the bodies of masonry,

5 feet above the level of the side-walk and tow-path, being 7 feet wide, leave 3 feet space for the passage of the pyramids. The ample width of the tow and foot-path is

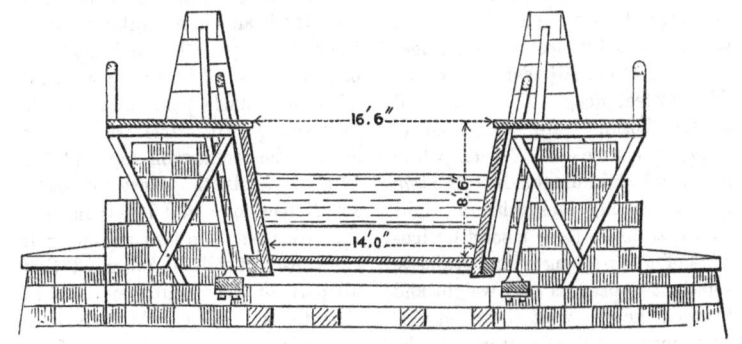

which are connected on each pier and abutment as bases for the pyramids which support the cables. These pyramids, which are constructed of three blocks of a durable, coarse, hard-grained sandstone, rise

therefore contracted on every pier; but this arrangement proves no inconvenience, and was necessary for the suspension of the cables next to the trunk.

small bridge on the Hertford and Ware Branch of the Northern and Eastern Railway, where it was necessary—in consequence of certain restrictions in the Act of Parliament authorising the construction of this branch—to construct a bridge for the purpose of carrying a common road over the River Lea, in the town of Ware, with a certain headway above the towing path, and yet not to raise the street more than a given amount. The span was to be 50 feet, and the conditions only admitted of a platform 18 or 20 inches in thickness. For this purpose a wrought-iron platform was designed, consisting simply of a series of cells, as shown in section in the following figure, the whole being of boiler plate, riveted together with angle-iron, as in our ordinary boiler building:—

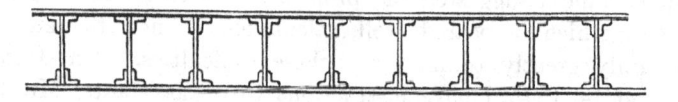

The bridge was not, however, carried out in conformity with the design. Instead of the platform consisting of wrought-iron plates, riveted together, forming one mass, it was constructed of separate wrought-iron girders, composed of wrought-iron plates riveted together, and arranged as in an ordinary cast-iron girder bridge.

It was reverting to this bridge that led me to apply wrought-iron with the view of obtaining a stiff platform to a suspension bridge, and the first form of its application was simply to carry out the principle described in the wooden suspended structure last spoken of, substituting for the vertical wooden trellis trussing, and the top and bottom cross braces, wrought-iron plates riveted together with angle iron. *The form which the idea now assumed was, consequently, simply a huge wrought-iron rectangular tube, so large that railway trains might pass through it, with suspension chains on each side.* The first arrangement, therefore, of the tubular structure was exactly similar in form to the trellis trussed wooden design before alluded to ; but it was evident that the action of the top and bottom of the tube, composed of thick wrought-iron plates, would be infinitely more efficient than the top and bottom braces, whose duty was chiefly to keep the side trusses in their vertical position. The top and bottom plates performed precisely the same duties as those of the top and bottom wheels of a common cast-iron girder. *It was now that I began to regard the tubular platform as a beam, and that the chains should be looked upon as auxiliaries.* The rectangular figure, although it admitted of great facilities for attaching chains, appeared ill suited for maintaining its form, and liable to become lozenge-shaped without a system of diagonal struts inside. This latter arrangement appeared impracticable, so long as the idea of the trains inside was adhered to.

The rectangular figure was also deemed objectionable, from the large surface which it presented to the wind — the side pressure due from a hurricane being very considerable. These circumstances suggested the propriety of circular or elliptical tubes, which appeared well calculated, if not to remove, certainly greatly to moderate these difficulties. On March 13 and 14, 1845, I gave instructions to my assistants, Mr. G. Berkley and Mr. W. P. Marshall, to prepare drawings of a tubular bridge in accordance with the last-mentioned views, the tubes being made with a double thickness of plate, top and bottom. All calculations were now made as to the strength of the tube, irrespective of the chains, by following the principle which had been adopted by Mr. Hodgkinson in determining an empirical formula for cast-iron girders. It will be seen hereafter that although this was not strictly correct reasoning, it was for practical purposes a near approximation to the truth; and as it disregarded the sides as an element of strength, it appeared to lead to unquestionably safe results. I could not at that time avail myself of the generally received theories of tubes, viz. that their strength was directly as their sectional area and depth, and inversely as their length, for no experiments had been then made confirmatory of such theory, or which furnished any data for practical purposes. The results, however, arrived at by assimilating the tube to a cast-iron girder, as has just been mentioned, were so favourable, that I determined at once to make use of this description of bridge, should the opposition to the design with two cast-iron arches, already described, prove formidable in our progress through Parliament with the Bill then pending.

At this juncture I was placed in a most difficult position.

Those interested in the navigation of the Menai Straits, as well as those who had, prior to this period, strenuously advocated Dynllaen in opposition to Holyhead as the proper terminus for a railway, had succeeded in inducing the Admiralty to give instructions to Sir John Rennie, Mr. J. M. Rendel, and Captain Vidal, to visit the Straits forthwith and report upon the probable injury which might accrue to the navigation of its waters by the erection of the proposed arches at the Britannia Rock.   I was too well acquainted with the overwhelming weight which is almost invariably given in such investigations to a long-established public interest, and the extreme jealousy with which any interference with it is watched, not to feel that the fate of my first design was sealed.   I stood, therefore, on the verge of a responsibility from which I confess I had nearly shrunk : the construction of a tubular beam of such gigantic dimensions, on a platform elevated and supported by chains at such a height, did at first present itself as a difficulty of a very formidable nature.   Reflection, however, satisfied me that the principles upon which the idea was founded were nothing more than an extension of those daily in use in the profession of the engineer. The method, however, of calculating the strength of the structure which I had adopted was of the simplest and most elementary character ; and whatever might be the form of the tube, the principle upon which the calculations were founded was equally applicable, and could not fail to lead to equally accurate results. When I began to regard the tube as a beam, one of the forms which the notion took was that of two huge double T cast-iron

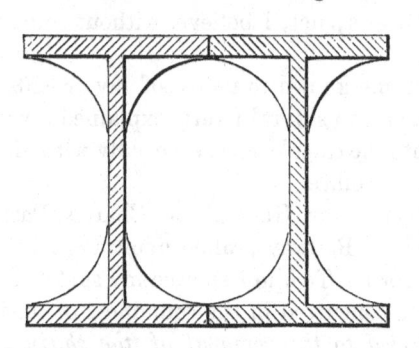

girders placed alongside of each other thus, sufficiently large for a railway between them.

In such a pair of beams, the area of the bottom section and the depth are the most important elements in calculating its strength. From this mode of treating the subject, it was obvious that the same view was strictly applicable to a tube whatever might be the form, the ultimate strength depending mainly on the area given to the top and bottom sections.

This was the shape which the subject took in my head between March 16 and 23, when Sir John Rennie, Mr. Rendel, and Captain Vidal, visited the site of the proposed bridge over the Menai Straits, in compliance with the instructions issued by the Lords Commissioners of the Admiralty, for the purpose of ascertaining how far the proposed cast-iron bridge of two arches was likely to interfere with the interests involved in the navigation. The reports of these gentlemen will be found annexed, from which it will be seen that the cast-iron bridge was deemed ineligible, and that a clear passage through the whole span, as in the existing chain of bridge, of at least 100 feet above high water line, would be insisted upon.

My apprehensions respecting the fate of the design for the cast-iron bridges were now realised, and it appeared evident that the tubular bridge was the only structure which combined the necessary strength and stability for a railway with the conditions deemed essential for the protection of the navigation.

It became my duty, then, to announce to the Directors of the Chester and Holyhead Railway Company that I was prepared to carry out a bridge of this description. They did me the honour of giving me their confidence after I had generally explained my views; not, I believe, without some misgivings on their part.

It soon became known to many of my friends what my intentions were, and to several I fully explained my views in detail, and entered into the calculations, especially with Mr. G. P. Bidder and others in my office.

General Pasley (now General Sir Charles Pasley), then Inspector-General of Railways, called upon me, I believe about the beginning of April. To him I showed my sketches and explained my views. He concurred in the soundness of the idea, *but most decidedly objected to the removal of the chains, urging as a reason, that no object could be answered by taking them down if once put in their place, for the purpose of contracting the tube.* To this argument I felt there was no sufficient answer, especially if any contrivance could be devised for making them serviceable in giving strength to the tube, the possibility of which I did not doubt, although from the observations I had made on the Stockton Suspension Railway Bridge, I considered there was considerable difficulty and several objections to rendering a flexible chain available for strengthening a rigid platform. General Pasley, however, urged so strongly the propriety of allowing them to remain on prudential grounds, that I ceased to urge the intention as a part of the design, for it was evidently a step which might be allowed to depend entirely on the results developed in the progress of the work. When I had explained to the directors my views, Mr. John Laird, the well-known iron ship builder, then one of the Chester and Holyhead Board, expressed his confidence in the great strength which such a structure as I proposed would possess, and adduced some instances where the extraordinary strength of iron ships had been tested when stranded. The most remarkable case, however, of this kind, where the strength of the hull of an iron vessel had been strikingly evinced, was brought before me by Mr. Miller, the eminent marine engine builder, and having occurred under his own eye, he was enabled to afford me the minutest information.

The incident here alluded to took place in launching the Prince of Wales iron steam vessel, at Blackwall, at the works of Messrs. Miller and Ravenhill, and was deemed so demonstrative of the excellence and surprising strength of iron ships, that an engraving was published by Mr. Miller, exhibiting the position and dimensions of the vessel, with a brief description of the accident as under.*

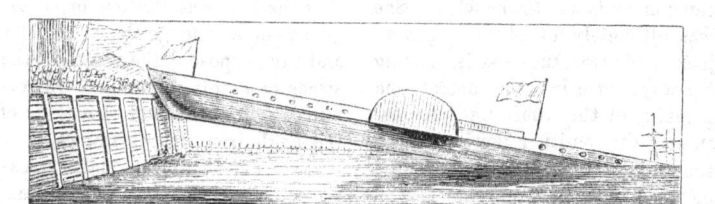

THE PRINCE OF WALES IRON STEAM VESSEL.

---

* This vessel is entirely of iron, and is intended for the Margate station. She is 180 feet long between the perpendiculars. In launching, the

The circumstances here brought to light were so confirmatory of the calculations I had made on the strength of tubular structures, that it greatly relieved my anxiety, and converted my confidence into a certainty that I had not undertaken an impracticable task.

The period was now approaching when I should be called upon to give evidence before a Parliamentary Committee on the subject of the proposed bridge. My late revered father, having always taken a deep interest in the various proposals which had been considered for carrying a railway across the Menai Straits, requested me to explain fully to him the views which had led me to suggest the use of a tube, and also the nature of the calculations I had made in reference to it. It was during this personal conference that Mr. William Fairbairn accidentally called upon me, to whom I also explained the principles of the structure I had proposed. He at once acquiesced in their truth, and expressed confidence in the feasibility of my project, giving me at the same time some facts relative to the remarkable strength of iron steam ships, and invited me to his works at Millwall to examine the construction of an iron steam ship which was then in progress. Mr. Fairbairn's experience in this department of engineering being well known to me, and also his investigations in connection with Mr. Hodgkinson on the subject of the strength of cast-iron, it occurred to me that he

----

cleet on the bow gave way in consequence of the bolts breaking, and let the vessel down, so that the bilge came in contact with the wharf. She was ultimately forced off by screw-jacks and two tug-vessels, cutting her way deeper into the concrete and planking of the wharf, until she assumed the position represented in the drawing; and at that period the distance measured from the face of the wharf to the point of contact of the vessel and the surface of the water was 110 feet. The whole of the deck in the centre of the vessel was left unfastened for the reception of the machinery. When completely afloat it was found that the shear of the vessel was not broken, and that she had received no injury, except that the bow was twisted in consequence of letting go the stern-rope, and thus exposing the vessel to the sweep of a strong ebb tide. On examination it was found that three of the angle iron ribs, or frames, were broken, and one of the plates cracked, occasioning a considerable leak, which was accompanied by no other inconvenience than that of filling the bow compartment as far as the first bulk-head; and after hauling the vessel into dock, the necessary repairs were effected in four days.

would be well qualified to assist me in the experimental enquiry which I had determined upon making prior to finally deciding on the exact dimensions of the tubes or mode of procedure.

He readily agreed to assist me, and it was forthwith decided to consider and arrange a series of experiments.

Nothing, however, but preparatory steps were taken, when the Bill for the deviation of the line in the vicinity of Bangor was brought before a Committee of the House of Commons on May 5, 1845.

The evidence I gave before the committee on the above day was received with much evident incredulity; so much so, that towards the end of that day's proceedings the committee stated they would require further evidence, and especially that of the Inspector-General of Railways, before they could pass the Bill authorising the erection of such a bridge as that which I had proposed. The preamble of the Bill was passed, but a resolution come to which left the question of the bridge entirely open for further consideration. In this position of things it became evident, from my general knowledge of the decided opinions held by the Inspector-General respecting the propriety of not dispensing with the chains, that I should not persist in the opinion that they were unnecessary. Accordingly it will be observed that whilst I expressed an unequivocal opinion as to the sufficiency of the tube alone, I was driven, from the circumstances that surrounded me, to leave the impression upon the minds of the committee that at all events the chains might be left as auxiliaries to the tubes if necessary.

The Bill passed the committee, and in due course became law by receiving the Royal assent, June 30, 1845.

I now commenced an experimental investigation on tubular constructions. The performance of the experiments I entrusted to Mr. Fairbairn. Before any experiments had been performed, he suggested a modification of my views, similar to that which has since been proposed by Mr. Cowper, as a mode of constructing rigid suspension bridges, and described in a paper read before the Institution of Mechanical Engineers, October 27, 1847.

The notion is the converse of the first, or beam platform. Mr. Fairbairn proposed to transfer the rigidity from the platform to the chain. This was effected by converting the chain

into a large flat tube or tubes, with sufficient flexibility to
assume a curved form, but with sufficient rigidity to resist much
distortion of curvature by unequal or varying pressure, such as
is usually communicated by the transit of a heavy weight along
the platform of suspension bridges.

The tubular construction in both ideas is resorted to for the
purpose of obtaining the requisite stiffness; and the question is
really, in which way are strength and stiffness attainable most
economically and efficiently.   This view occupied my attention
for some little time; but the difficulties of erection appeared to
me insuperable, whereas, with the rigid platform, the ordinary
chains offered great facilities for constructing and erecting the
tubes.   The rigid beam, instead of the rigid chain, was therefore
persevered in as preferable, not only because it afforded greater
facilities for erection, but on account of the rigidity of the
curved tube being very problematical.

There can, however, be no question that a rigid suspension
bridge, with a tubular chain, in some cases, may be employed
with success.   Its construction, however, did not appear to offer
such advantages as to justify my rejecting the rigid tube, which
offered greater facilities for erection, as also the probability of
dispensing with the chains altogether.

We had not proceeded far in this experimental investigation
when Mr. Fairbairn suggested that Mr. Eaton Hodgkinson's aid
should be solicited.   To this proposal I instantly consented;
for being familiar with the valuable contributions of this gentleman to engineering science, more especially in the department
which comprehends the very subject then engrossing my attention, I felt, considering the responsibility which I had publicly
assumed, that I should be doing injustice to the Board of
Directors, who had placed such confidence in me, if I did not
avail myself of all the practical and scientific aid which might
be within my reach.   I did so, freely and unhesitatingly, from
every quarter.   The experiments conducted by these gentlemen
are given in detail in a subsequent chapter.   The facts elicited
by these experiments were carefully discussed with them from
time to time, and the best method of pursuing the investigation
was determined upon.   In the majority of the early experiments
failure took place by the crushing of the thin plates in the upper
side of the tubes.   This defect induced me at once to return to

the original form in which the tube had occurred to me, viz. that of an ordinary flanged girder. To carry out this notion, keeping in view the tendency of the plates to buckle, a double top of corrugated iron was applied, and a corresponding increase made in the strength of the bottom was attended with favourable results.

The top of the tube thus came to be considered simply as a series of parallel hollow pillars to resist the compression to which it was subjected by transverse strain. Among the numerous modifications which presented themselves, a series of rectangular cells possessed so many practical advantages, as regards construction, that I did not hesitate to give this arrangement a preference; and although subsequently some advantages in the use of circular cells were clearly developed by the experiments, I did not consider them of sufficient importance to change my decision. There is no doubt, from subsequent experience, that the fears at this time entertained with respect to buckling were, to a considerable extent, exaggerated, for the thickness of plates is an element of far greater importance in this resistance to buckling than was then imagined.

In March 1846, sufficient data were accumulated to enable me, for the first time, to decide somewhat definitely on the required dimensions, and I accordingly gave instructions to my assistant, Mr. Edwin Clark, to prepare a model of the tube, which was made with rectangular cells in the top and bottom, the sectional area of the top being then provisionally fixed at 600, and that of the bottom at 400 square inches. The various modifications which afterwards took place, as new facts were disclosed, will be found fully described hereafter.

The success which has attended our exertions in this laborious and anxious investigation demands of me this public acknowledgement.

To Mr. Fairbairn I am indebted for the zeal with which he entered upon the experimental investigation, for the confidence he displayed in the success of my design, for the sound practical information which he brought to bear on the subject, and the assistance he rendered as we progressed.

To Mr. Hodgkinson for devising and carrying out a series of experiments which terminated in establishing the laws that regulate the strength of tubular structures in a manner so satisfactory that I was enabled to proceed with more confidence than I otherwise should have done.

To Mr. Edwin Clark, the resident engineer, for the important assistance he rendered me in strictly scrutinising the results of every experiment, whether made by Mr. Fairbairn or by Mr. Hodgkinson, and for the separate and independent scientific analysis to which they were invariably subjected by him before I finally decided upon the form and dimensions of the structure or upon any mode of procedure.

I have now brought the history of the Conway and Britannia Bridges to a date after which all that was done has either been communicated in official reports to the Board of Directors, or has been carefully registered by my assistant, Mr. Edwin Clark, who has given his undivided attention to the subject since the beginning of 1846, during which period he has collected a mass of information which cannot fail to be both interesting and important to the profession to which he belongs. I cannot close this statement respecting two works which have caused me years of increasing and intense anxiety without expressing my regret that one of the gentlemen to whom I have always been most anxious to award all credit to which he is entitled should have endeavoured to enhance his own claims by detracting from the credit fairly due to all those with whom he has been associated in this great work. But I sincerely trust the facts and views put on record in the following pages will enable those who take an interest in the subject to do justice to all parties concerned.

x

INDEX.

INDEX.

ABB

A BBAS Pasha, Viceroy of Egypt, summons Robert Stephenson to Cairo, ii. 173

Abyssinia, the viceroy of Egypt in, ii. 247

Adams, Mr. Bridge, on the mechanical defects of railways, i. 290

Addison, Dr., at Tommy Rutter's school at Long Benton, i. 20
— his death, i. 20

Aire, Robert Stephenson's iron bridge over the, ii. 68

Airey, Professor, on the Royal Gauge Commission, ii. 12

Albert Bridge, the Royal, at Saltash, ii. 68

Alexandria, Robert Stephenson at, ii. 174
— the railway from Alexandria to Cairo, ii. 174

Algiers in 1857, ii. 247

America, Lakes of, flowing into the river St. Lawrence, ii. 191

Anger, Mr., in Egypt, ii. 179, note

Anglesea, Island of, its distance from the mainland of Carnarvonshire, ii. 74

Arches, invention of, in bridges, ii. 32
— Egyptian, ii. 32
— Mr. Dixon's method of fixing, ii. 289

Army Estimates of 1856, Robert Stephenson's speech on the, ii. 147

Arnold, Dr., Robert Stephenson's interview with, i. 202

Athenæum Club, Robert Stephenson at the, ii. 230

Atlantic and St. Lawrence Railway, formation of the, ii. 192

Atlantic telegraph, proposition for an, ii. 250.

ATM

Atmospheric system of railway propulsion, history of the, i. 292
— Denys Papin's notions of atmospheric propulsion, i. 299
— George Medhurst's inventions, i. 299
— Mr. John Vallance's tubes and experiments, i. 301
— Mr. Henry Pinkus's patent, i. 301
— Messrs. Clegg and Samuda's patents, i. 302
— first trial of the system on the Birmingham, Bristol, and Thames Junction Railway, i. 303, 338
— description of the apparatus as given by the inventors, i. 303
— Mr. James Pim's advocacy of the atmospheric system, i. 309
— favourable report of Sir F. Smith and Professor Barlow, i. 310
— trial of the system on the Kingston and Dalkey Railway, i. 310, 338
— details of the apparatus, i. 311
— summary of the chief arguments in favour of the atmospheric system, i. 312
— Robert Stephenson's report to the Chester and Holyhead Directors, i. 316
— Mr. Samuda's calculation of the expense of the tubes and engines per mile, i. 324
— interest of the public in the atmospheric question, i. 329
— the Croydon and Epsom Railway, i. 330
— appointment of a Committee of the House of Commons to inquire into the merits of the atmospheric plans, i. 331
— their report in its favour, i. 332

ATM

Atmospheric system—*continued*
— culminating point of the history of the atmospheric pipe, i. 337
— the five cases of the actual practice of the atmospheric system, i. 338
— details of the tubes used between Forest Hill and Epsom, i. 340
— the South Devon Railway, i. 346
— history of the trial of the atmospheric system in France, i. 350
— determination to abandon it in that country, i. 354
— remarks on the results of the trials made of the atmospheric plan, i. 356
— mechanical efficiency of the system, i. 356
— question of economy, i. 358
— and of the general applicability of the system to railway traffic, i. 358
Avalanche galleries of the Alpine roads, ii. 79

BAINES, Mr. Edward, on the Alexandria and Cairo Railway, ii. 179, *note*
Ballast engine, the first, on the banks of the Tyne, i. 6
Barlow, Professor, his favourable report on the atmospheric system, i. 310
— on the Royal Gauge Commission, ii. 12
Beam, return to the form of the, in iron bridges, ii. 41
— advantages of the beam or girder system over the arch or suspension forms, ii. 43
— the earliest iron beams used in building, ii. 44
— Mr. Telford's 'Essay on the Strength of Cast Iron,' ii. 44
— Mr. Rastrick's girders at the British Museum, ii. 44
— bowstring girders, ii. 46, 65
— trussed girders, ii. 47, 65
— first application of hollow wrought iron girders to bridge construction, ii. 62, 63
— details of girder bridges, ii. 64
Beds, colliers', i. 9
Belgium, railways in, i. 220
— Robert Stephenson's account of railway progress in, ii. 184

BIR

Bell, John and Henry, said to have been the discoverers of the blastpipe, i. 145
Benha, railway tubular bridge over the Nile at, ii. 176
— the designs made by Mr. G. R. Stephenson, ii. 179, *note*
Bentinck, Lord George, his proposal to subsidise Irish Railway Companies at the time of the famine of 1846, ii. 131
— consults Robert Stephenson, George Hudson, and Mr. Laing, ii. 132
— his speech in the House of Commons on the subject, ii. 133
Benton, Long, George Stephenson at, i. 10, 12
— the 'street' of, i. 10
— Sweaton's atmospheric engine at, i. 12, *note*
— road from Newcastle to, i. 13
— Tommy Rutter's school at, i. 19, 20
— Antony Wigham of, i. 26
— visit of Robert Stephenson to, in 1857, ii. 237
Benton Banks, i. 13
Betts, Edward Ladd, one of the contractors for building the Victoria tubular bridge over the St. Lawrence, ii. 187, 209
Bidder, Mr. George Parker, commencement of his friendship with Robert Stephenson, i. 60
— his opposition to the atmospheric system of propulsion, i. 334
— accompanies Robert Stephenson on a trip to Norway, ii. 130
— his report on the state of the public work in Ireland in 1846, ii. 132
— joins Robert Stephenson on a yachting excursion, ii. 241.
— completes the Norwegian railway, ii. 244
Bidder, Mr., jun., in Egypt, ii. 179, *note*
Bidder, Mr. Samuel, his account of a trip to Canada with Robert Stephenson in 1853, ii. 180
Bigge, Mr. Matthew, accompanies Robert Stephenson on a visit to the North, ii. 237
Birmingham, Bristol, and Thames Junction Railway, application of the atmospheric system to the, i. 303, 338

BIR

Birmingham and Gloucester Railway, ii. 9

Birket - el - Saba, railway tubular bridge over the Nile at, ii. 176

— the designs made by Mr. G. R. Stephenson, ii. 179, *note*

Blackwall Railway, Mr. G. P. Bidder's statement respecting the mode of working the, i. 328

Blast-pipe, history of the, i. 144

— Mr. Hedley Wylam's mode of letting off the waste steam, i. 144

— George Stephenson's 'Puffing Billy,' i. 145

— the Bells and James Stephenson, i. 145, 146

— George Stephenson's letter to Mr. Phipps, i. 148

Blenkinsop's patent locomotives, i. 264

Blisworth cutting, cost of the, i. 194

Bogie engines on North American railways, ii. 180

Boiler tubes, locomotive, of the Messrs. James, i. 51, 52

— the multitubular boilers of Mr. Henry Booth, i. 52

— the multitubular, in locomotive engines, i. 118, 119, 126–128

— 'the Rocket,' the first engine with multitubular boilers, i. 144

Bonomi, Mr. Joseph, his paper on an enormous granite sarcophagus, ii. 250

Booth, Mr., treasurer of the Liverpool and Manchester Railway, his invention of the multitubular boiler, i. 52, 119

— his account of the question of locomotives against stationary engines, i. 123, *note*

Borthwick, Mr. M. A., in Egypt, ii. 179, *note*

Boulton and Watt, Messrs., their use of iron beams in building, ii. 44

Boussingault, M., visits Robert Stephenson at Santa Ana, i. 91, 92

Bowstring girders, ii. 46, 65

Boyne, lattice girder bridge over the, ii. 67

Boxer, Mr. F. N., his description of the Victoria bridge over the river St. Lawrence, ii. 190, *note*

Braithwaite and Erichson, Messrs., their engine, 'The Novelty,' i. 142, 143

BRI

Braithwaite, Mr., appointed engineer of the London and Colchester Railway, i. 238

Brandling, Mr., his mansion, i. 14

— his kindness to George Stephenson, ii. 143

Brandreth, Mr., his ingenious engine, 'The Cyclops,' i. 142

Brassey, Thomas, one of the contractors for constructing the Victoria viaduct over the St. Lawrence, ii. 187, 209

Breakwater at La Guayra, Robert Stephenson's report against the construction of a, i. 78

Bridges, iron, Mr. Stephenson's large practice in iron bridges, ii. 30

— the Britannia, High-Level, and Victoria bridges, ii. 30

— remarks on the application generally to iron bridge building, ii. 31

— the first iron arch bridges, ii. 33

— Dr. Desaguliers, ii. 33

— bridge across the Severn at Coalbrookdale, ii. 33

— Mr. Telford's bridge over the Severn, ii. 34

— arch iron bridge over the Wear, ii. 34

— Tom Paine's bridge, ii. 34, 35

— bridge at the Wearmouth ferry, ii. 34

— and over the Thames at Staines, ii. 35

— iron bridges in France, ii. 36

— the Ponts du Louvre and d'Austerlitz, ii. 36

— Vauxhall and Southwark bridges, ii. 36

— cast and wrought iron for bridges, ii. 36, 37

— properties of wrought iron as distinguished from those of cast iron, ii. 37

— suspension bridges, ii. 37

— the foot-bridge across the Tees, at Middleton, ii. 37

— Captain (afterwards Sir Samuel) Brown's suspension bridges, ii. 38

— Mr. Telford's suspension bridges over the Menai Straits, ii. 39

— introduction of railways, and consequent large demands for iron bridges, ii. 39, 40

— return to the form of the simple beam bridge, ii. 41

— comparison of the advantages of

BRI

Bridges—*continued*

the three different systems of iron bridges, ii. 42

— advantages of the beam or girder over the other systems, ii. 43.

— Mr. Hodgkinson's improvements in girder bridges, ii. 45

— the bowstring girder, ii. 46

— the trussed girder, ii. 47

— the bridges over the river Lea and over the Minories, ii. 48

— accident to the bridge over the Dee at Chester, ii. 48

— Robert Stephenson's description of the defects of compound girders, ii. 53

— his evidence before the Iron Railway Structure Commission, ii. 55

— iron bridges of the Trent Valley Railway, ii. 55

— and of the Florence and Leghorn line, ii. 55

— circular issued by the Commissioners of Railways, ii. 55

— minutes published also by them, ii. 56

— commission appointed by royal warrant to investigate the subject of iron bridges, ii. 56, 57

— evidence taken before the Commissioners, ii. 57, 58

— Professors Willis and Stokes's paper on the deflection of beams under moving loads, ii. 58

— report of the Commissioners to Her Majesty, ii. 58

— first application of wrought-iron girders for bridge construction, ii. 62

— general view of the state now attained in the art of iron bridge building, ii. 64

— three classes of iron bridges, ii. 64

— details of girder bridges, ii. 64

— Mr. Stephenson's bridges over the Aire, over the Nile, and over the Wear, ii. 68–70

— account of the Britannia bridge, ii. 73 *et seq.*

— Mr. Telford's suspension bridge over the Straits, ii. 77

— first idea of the tubular combination, ii. 82

— novelty of such constructions, ii. 84

— experiments for the Britannia bridge, ii. 85

BRI

Bridges—*continued*

— important principles derived from these experiments, ii. 86

— description of the Britannia tubular bridge, ii. 100

— and of the Conway bridge, ii. 109

— account of the High-Level bridge at Newcastle-on-Tyne, ii. 113

— description of this bridge, ii. 117

— Mr. Stephenson's experiments upon different kinds and different mixtures of cast iron, ii. 126

— the Kaffre-Azzayat tubular bridge over the Nile, ii. 174–176

— Robert Stephenson's description of it, ii. 177

— the Victoria viaduct over the river St. Lawrence, ii. 180, 186

— detailed account of this structure, ii. 190 *et seq.*

— Robert Stephenson's article in the 'Encyclopædia Britannica' on, ii. 31

Bridge, swing, over the Nile, ii. 220

Bright, Mr. John, M.P., i. 257

Brighton, chain pier at, ii. 38

Bristol and Gloucester Railway Extension, ii. 9

— the first 'break of gauge' on the, ii. 9

Britain, Great, expense of railways in, as compared with those of foreign countries, i. 289

Britannia bridge, account of the, ii. 73, 285

— Mr. Stephenson's selection of the site of the bridge, ii. 80

— first idea of the tubular combination, ii. 80

— the Britannia Bridge Bill passed, ii. 84

— Mr. Fairbairn and Mr. Hodgkinson's experiments, ii. 85

— important principles derived from the experiments, ii. 86

— Mr. Stephenson's report, ii. 86

— buckling or collapse of the plates of the top part, ii. 87

— laying the first stone of the bridge, ii. 89

— the large model of the bridge prepared by Mr. Fairbairn, ii. 90

— results of experiments with this model, ii. 91

— means of placing the tubes in their positions, ii. 91

BRI

Britannia bridge—*continued*
— manufacture of the tubes, ii. 94
— floating and hoisting them up, ii. 96
— completion and opening of the bridge, ii. 100
— description of the structure, ii. 100
— principle of continuity, ii. 100
— the tubes, ii. 101
— Mr. Stephenson's explanations of peculiarities in the combinations, ii. 104
— the towers and abutments, ii. 107
— architectural design, ii. 108
— cost of the work, ii. 108
— its durability, ii. 109
— visit to the bridge in 1857, ii. 242

Britannia rock, in the Menai Straits, ii. 80

Broughton, Robert Edwards, his death, ii. 252

Brown, Mr., his vacuum-engines, i. 71

Brown, Captain (afterwards Sir Samuel), his chain cables, ii. 38
— his suspension bridges, ii. 38, 76
— his chain pier at Brighton, ii. 38

Bruce, Mr., his academy in Newcastle, i. 34
— Robert Stephenson a pupil of his, i. 34

Bruce, Mr. George B., and the Royal Border bridges, ii. 140

Brunel, Mr. I. K., appointed engineer of the Great Western Railway, i. 238
— his evidence in favour of the atmospheric system of propulsion, i. 332
— his friendship for Robert Stephenson, ii. 2
— his proposal of the broad gauge for the Great Western Railway, ii. 3
— advantages anticipated by him from a broad gauge, ii. 4
— his mode of treating the anticipated inconvenience of 'break of gauge,' ii. 5
— his report to the Great Western directors in 1838, ii. 5
— his theory of railway districts, ii. 8
— his inconsistencies, ii. 12
— failure of his theory of railway districts, ii. 13
— his expedients for meeting the

CAN

Brunel, Mr. I. K.—*continued*
evils of the 'break of gauge,' ii. 14
— his indifference to pecuniary expense, ii. 15
— establishment of the Railway Clearing House, ii. 16
— Robert Stephenson's remarks on Brunel's loose box system, ii. 22
— his argument of competition between the different lines, ii. 26
— his evidence on iron railway structures, ii. 58
— his iron bridge over the Thames near Windsor, ii. 65
— his iron bridges over the Wye at Chepstow and over the Tamar at Saltash, ii. 67
— his assistance in raising the tubes of the Britannia bridge, ii. 98
— his Leviathan ship, ii. 244
— his visit to Egypt, ii. 249
— his report to the Great Western railway on the gauge question, ii. 269
— Sir Isambard, his plan for constructing brick arches without centering, ii. 288

Brunton, Mr., of Neath Abbey Works, i. 70

Brussels, Robert Stephenson's visit to, i. 114

Brussels and Ghent Railway, opening of the, i. 220

Burn, Ann, of Walsingham, Robert Stephenson's visits to, i. 18

Burstall, Mr., of Leith, his locomotive, the 'Perseverance,' i. 142, 144

CABLES, chain, first introduction of, into the navy, ii. 38

Cabmen, Robert Stephenson's treatment of, ii. 165

Calisthenic exercises for ladies, i. 104

Callerton, Black, George Stephenson at, i. 3

Canada in 1827, i. 110
— Robert Stephenson's trip to, ii. 180
— — his speech at Montreal on railways in, ii. 182
— inland lake navigation of Canada, ii. 192
— first efforts in railways in, ii. 192
— the Atlantic and St. Lawrence Railway, ii. 192

CAN

Canada—*continuod*
— the Quebec and Richmond and Montreal and Kingston Railways, ii. 192
— the Grand Trunk Railway of, ii. 193
— main objects of the railway system of, ii. 192, 193-4
— extent and importance of the Grand Trunk Railway, ii. 194
— Robert Stephenson on railways and their importance in Canada, ii. 207
Cannibalism at sea, i. 106, *note*
Canterbury and Whitstable Railway, opening of the, i. 154
Capper, Mr. Charles, anecdote of him at the dinner at Dunchurch, i. 212
Carr, Mabel, married to Old Robert Stephenson, i. 2
— their children, i. 2
Charlton, John, engaged to Fanny Henderson, i. 5
— his death, i. 5, 8
Chat Moss, George Stephenson's difficulties at, i. 112
Cheffins, Mr., jun., assistant surveyor to the Alexandria and Cairo Railway, ii. 179, *note*
Chester and Holyhead Railway:—
— rejection of a bill for erecting a bridge of two arches over the Straits, ii. 81
— the tubular Britannia bridge, ii. 82
— the bill for the tubular bridge passed, ii. 84
— Robert Stephenson's examination of the atmospheric system in reference to this line, i. 315
— fatal accident at the Dee bridge, ii. 48
— engineering interests and importance of this line, ii. 78
— course of the line, ii. 78, 79
— the bill passed with a hiatus of five miles, ii. 80
Chien, or *Hund*, the, i. 99
Circumferentor, or mining compass, made by Robert Stephenson, i. 48
Clanny, Dr., his safety-lamp, i. 37
Clark, Mr. Edwin, his evidence on iron bridges, ii. 58
— Robert Stephenson's letters to, from Alexandria and Algiers, ii. 246

CON

Clay Cross collieries, George Stephenson's property in the, i. 240
Claxton, Captain, R.N., his assistance in raising the tubes of the Britannia bridge, ii. 97
Clearing House, Railway, establishment of the, ii. 16
— its leading principles, ii. 16
— its returns for 1845, ii. 17
Clegg, Mr. Samuel, his patent atmospheric tubes, i. 302
— associates himself with Messrs. Samuda, i. 302
— trial of his plan on the Birmingham, Bristol, and Thames Junction Railway, i. 303
Clocks, colliers', i. 9
Coaches *versus* Railways, i. 170, 171
Coalbrookdale, iron bridge over the Severn at, ii. 33
Cobden, Mr., his motion respecting a uniform gauge, ii. 11
Cogyn river, viaduct over the, ii. 79
Coining press, invented by Robert Stephenson, i. 71
Collieries near Swannington, i. 164
— the Snibston estate, i. 164
— at Medomsley, i. 215
— George Stephenson's mines at Clay Cross, i. 240
— tramways in 1676, ii. 3
Colliers, their beds and clocks, i. 9
Colombia, condition of, in 1827, i. 103, 109
Colombian Mining Association, the, i. 66
— proposition to Robert Stephenson to accompany the expedition, i. 68
— agrees to go, i. 72
— found to be a losing concern, i. 93, 94
— disappointment of the directors, i. 98
— the secretary's blunder, i. 99
— condition of the Company's affairs at the end of Stephenson's term, i. 101
— machinery constructed for the, i. 130
Commercial Road, the tramway laid down by Mr. James Walker in the, i. 227
Commercial Road Railway, i. 228
Constantinople, winter of 1858 in, ii. 246
Contractors, Railway, in 1833, i. 189, 193

CON

Conway river, railway iron bridge over the, ii. 78

Conway bridge, description of the structure, ii. 109

Cooke, Mr. William, and the electric telegraph, i. 235

Cornish mining, Robert Stephenson's report on, i. 69

— miners, in Colombia, i. 87

— their demoralisation, i. 87

— their riots in Robert Stephenson's house, i. 89, 90

— athletic sports with them, i. 91

Croydon and Epsom Railway, adoption of the atmospheric system on the, i. 330, 341

Crumlin, the iron triangular framed girder bridge at, ii. 66

Cubitt, Mr. (afterwards Sir W.), his evidence in favour of the atmospheric system of propulsion, i. 332

Cundy, Mr., his proposed railway from London to Brighton, i. 222, 223

DALHOUSIE, Lord, his motion for a commission for deciding on a uniform gauge, ii. 11

Darien, Isthmus of, roads in the, i. 95

Davy, Sir Humphry, invention of his safety-lamp, i. 37

— the safety-lamp controversy, i. 37–39

— reward awarded to Sir Humphry, i. 39

Dean river, i. 13

Dee bridge, accident to the, ii. 48

— mode of strengthening it, ii. 55

Dixon, Mr., his method of fixing arches, ii. 289

Dockray, Mr. R. B., his account of Mr. Dixon's method of fixing arches, ii. 289, note

Dodd, Mr. R. B., his proposal for a high-level bridge at Newcastle-on-Tyne, ii. 114

Dolly Pit, George Stephenson at the, i. 3

Dripsey Paper Works, the two Stephensons at the, i. 63

Dröbak, the 'Wenham Lake ice' obtained at, ii. 258

Dublin compared with Edinburgh, i. 63

Duff, Mr., sub-assistant engineer in Egypt, ii. 179, note

Dynllaen, Port, ii. 77

EYR

EARTHQUAKES, effect of, on the town of Mariquita, i. 83

Eau Brink Commissioners, their dispute with the Norfolk Estuary Company, ii. 137

Ebrington, Lord, ii. 146

Edinburgh university, Robert Stephenson's residence in the, i. 55

Egypt, Robert Stephenson in, ii. 173

— introduction of railway communication into, ii. 173

— the great pontoon over the Nile, ii. 173

— Robert Stephenson's trip to, in 1858, ii. 244

— the air of the desert, ii. 247

— Mr. Bonomi's paper on an enormous granite sarcophagus, ii. 250.

— letter from the Pasha of Egypt to Robert Stephenson, ii. 261

Egyptian arches, ii. 32

Electric telegraph, invention of the, i. 235

Elliot, Hannah, of Ryle, Robert Stephenson's visits to, i. 19

Ellis, Mr., of Liverpool, i. 75

Empson, Charles, accompanies Robert Stephenson to Carthagena, i. 103

— and to England, i. 111, 113

'Encyclopædia Britannica,' Robert Stephenson's article on iron bridges in the, ii. 31

— Robert Stephenson's article on 'Iron Bridges' quoted, ii. 177

Engine, the first ballast, on the banks of the Tyne, i. 6

Engines. See Locomotive engines

Engine-drivers in 1828, i. 115

Euston Square, the stationary engines and ropes originally used at, and Camden Town, i. 209

Evans, Mr., his contract for the Conway bridge, ii. 111

Exhibition of 1851, Robert Stephenson on the committee, ii. 130

— Colonel Sibthorp's speech and motion against the proposal for the, ii. 146

— Robert Stephenson elected a member of the committee, ii. 166

— his offer of a loan of 1,000l. to the committee, ii. 166, note

Eyre Arms Hotel, the draughtsmen's room at the, i. 191

FAI

FAIRBAIRN, Mr. W., services rendered by him to Robert Stephenson, ii. 299, 303
— his evidence on iron railway structures, ii. 58
— takes out a patent for hollow wrought-iron girders, ii. 63
— his bridges on this plan at Blackburn and Bolton, ii. 63
— his experiments for the Britannia bridge, ii. 85
— prepares the large model of the bridge at Millwall, ii. 90
Ferry, steam, on the Nile, ii. 174
— Mr. Sopwith's account of it, ii. 174
Flatchat, M. Eugène, his information on the trial of the atmospheric system in France, i. 350 *et seq.*
Florence and Leghorn Railway, iron bridges of the, ii. 55
France, iron bridges in, ii. 33, 36
— Robert Stephenson's account of railway progress in, ii. 184
Freemasons' Tavern, Royal Society Club dinners at the, ii. 231
Friar's Goose pumping engine of George Stephenson, i. 51
Fowler, Mr., in Egypt, ii. 219, *note*
Fox, Mr. (now Sir C.), his evidence on iron railway structures, ii. 58
— his iron bridges, ii. 65

GAUGES, the battle of the, ii. 1
— the Great Western Railway in 1833, ii. 2
— Brunel's proposal of the broad gauge for the Great Western, ii. 3
— the 4-feet 8½-inch gauge, ii. 3
— gauge of George Stephenson's first public railway, ii. 3
— gauges of some of the tramways in the mineral districts, ii. 4
— advantages anticipated by Brunel from a broad gauge, ii. 4
— obvious objections to a wide gauge at that period of railway history, ii. 4
— Brunel's first position, ii. 5
— — his mode of treating the inconvenience of 'break of gauge,' ii. 5
— — his examination before the Gauge Commissioners in 1845, ii. 6

GER

Gauges—*continued*
— Brunel's theory of railway districts, ii. 8
— the Great Western Railway constructed, ii. 8
— first 'break of gauge,' ii. 9
— goods traffic the grand cause of difficulty in 'break of gauge,' ii. 10
— gauge pamphleteers, ii. 10
— the Oxford and Wolverhampton contest, ii. 10
— motions of Lord Dalhousie and Mr. Cobden, ii. 11
— appointment of the Royal Gauge Commission, ii. 11
— Brunel's inconsistencies, ii. 12, 13
— classified list of witnesses examined by the Gauge Commissioners in 1845, ii. 17, *note*
— Robert Stephenson's evidence on the question of the, broad and narrow, ii. 21
— — his objections to the system of double, ii. 23, 25
— table of railway lines completed, in progress, and projected, on the broad and narrow gauge plans, in 1845, ii. 24
— report of the Gauge Commissioners, ii. 25
— the 'Dialogues of the Gauges' in the 'Railway Record,' ii. 26
— Brunel's argument of competition between the different lines, ii. 26
— picture from the 'Illustrated News' of the troubles consequent on a break of gauge, ii. 27
— passing of the Act for Regulating the Gauge of Railways, ii. 28
— Brunel's report to the Great Western Railway Company on the question of gauge, ii. 269
— Robert Stephenson's report, ii. 272
Gay, Mr. E. H., of Pennsylvania, his survey for a bridge across the St. Lawrence, ii. 198
Geographical Society, their dinners, ii. 230
— discussion on the Suez Canal question at the, ii. 250
'Geordie' safety-lamp, the, i. 37
— anecdote of George Stephenson's trial of the, i. 38
Germany, erection of iron bridges in, ii. 34

GIB

Gibbons, Mr. Barry, his evidence in favour of the atmospheric system of propulsion, i. 331

Gibbs, Mr. Joseph, his proposed railway between London and Brighton, i. 222

Giles, Mr. Francis, appointed engineer of the London and Southampton Railway, i. 238

Girders, bowstring, ii. 46, 65
— trussed, ii. 46, 65
— Robert Stephenson's description of the defects of compound girders, ii. 53
— first application of wrought iron, to bridge construction, ii. 62, 63
— details of girder bridges, ii. 64
— the I-shaped girder, ii. 65
— examples of other, ii. 66, 67

Glyn, Mr., his part in the formation of the Railway Clearing House, ii. 16

Gooch, Mr. T. L., Robert Stephenson's letter from Alexandria to, ii. 245

Graham, Mr., surveyor to the Alexandria and Cairo Railway, ii. 179, note

Grand Junction Canal, rights and privileges of the, reserved by the London and Birmingham Railway Act, i. 196
— temporary railway bridge over the canal removed by the Canal Company's engineers, i. 198, 199
— the Canal Company restrained by the Master of the Rolls, i. 200
— the bowstring girder bridge over the, at Weedon, ii. 47

Grand Junction Railway, projection and execution of the, i. 239

Grand Trunk Railway Company of Canada, formation of the, ii. 193
— its extent and importance, ii. 194

Great Eastern ship, Mr. Brunel's, ii. 244

Great George Street, offices of Robert Stephenson's friends at No. 24, ii. 246

Great Western Railway Company, report of Mr. Brunel on the question of guage, ii. 269

Great Western Railway, Mr. Brunel appointed engineer of the, i. 238
— proposal with respect to it in 1833, ii. 2
— defeat of the first bill, ii. 2

HIG

Great Western Railway—continued
— Brunel's proposal of the broad gauge, ii. 3
— the line authorised and constructed on the broad gauge, ii. 8
— its triumph over the London and Birmingham, ii. 11

Green, Mr. John, his proposal of a high-level bridge at Newcastle-on-Tyne, ii. 114

Greenwich Observatory, Robert Stephenson at the, ii. 251

Gosforth Hall, i. 14

HACKWORTH, Mr., his engine, the 'Sanspareil,' i. 142, 143

Halhette, M., his improvements in atmospheric tubes, i. 350

Hammer, throwing the, i. 42

Hardcastle, Mr., in Egypt, ii. 179, note

Harrison, Mr. Thomas E., his drawings of the High-Level bridge over the Tyne at Newcastle, ii. 117
— and the Royal Border bridge, ii. 140

Harrison, Mr. Joseph, in Egypt, ii. 179, note

Hawthorn, Robert, i. 6
— erects the first ballast engine on the banks of the Tyne, i. 6
— his intimacy with George Stephenson, i. 9, 23

Hayter, Sir William, testimonial to, ii. 156

Henderson, Ann, i. 3
— refuses George Stephenson, i. 4
— godmother to Robert Stephenson, i. 7

Henderson, Fanny, i. 4
— her character, i. 5
— her early life, i. 5
— her marriage to Geo. Stephenson, i. 5
— her death, i. 15
— Robert Stephenson's visits to her grave, ii. 244

Hertford and Ware branch railway, Robert Stephenson's small bridge on the, ii. 294

Hetton Colliery Railway, begun and finished by George Stephenson, i. 49

High-Level bridge at Newcastle-on-Tyne, ii. 113
— object of the bridge, ii. 113

HIG

High-Level bridge—*continued*
— ravine of the Tyne, ii. 113
— the ancient low-level bridge, ii. 113
— early proposals for a high-level bridge, ii. 114
— Mr. Green's scheme, ii. 114
— formation of the High-Level Bridge Company, ii. 114
— taken up by the promoters of the Newcastle and Berwick Railway, ii. 116
— passing of the bill, ii. 117
— Mr. Harrison's drawings, ii. 117
— description of the bridge, ii. 117
— the piers, ii. 118,
— the iron superstructure, ii. 119
— contract for the building, ii. 124
— erection and completion of the bridge, ii. 127
Hill, Mr. (now Sir) Rowland, at the Royal Society Club, ii. 252
Hinde, Mr. John Hodgson, his promotion of the Newcastle High-Level bridge, ii. 114
Hindmarsh, Elizabeth, second wife of George Stephenson, i. 3
— married to George Stephenson, i. 50
Hodges, Mr. James, his description of the Victoria bridge over the river St. Lawrence, ii. 190, *note*
Hodgkinson, Mr. Eaton, his evidence on iron railway structures, ii. 58
— his experiments for the Britannia bridge, ii. 85, 90
— his death, ii. 265
Holyhead trunk road, construction of, by Telford, ii. 74, 76
Holyhead, importance of the port of, ii. 74
— decided on as the point of departure for Ireland, ii. 78
Honda, scenery of the road from Mariquita to, i. 83
Howick, Lord, his motion for a committee to examine the merit of the atmospheric system, i. 331
Hudson, Mr. George, chairman of the Newcastle and Darlington Railway, i. 251
— his leading characteristics, i. 252
— presides at the Newcastle and Darlington Railway banquet at Newcastle, i. 257
— contrasted with Robert Stephenson, i. 261

JAC

Hudson, Mr. G.—*continued*
— his part in the formation of the Railway Clearing House, ii. 16
— becomes vice-chairman of the High-Level Bridge Company, ii. 114
— consulted by Lord G. Bentinck as to railway works in Ireland, ii. 132
Hull docks, Robert Stephenson's consultation with Smeaton about the, i. 254
Huskisson, Mr., his inspection of the Liverpool and Manchester Railway in 1829, i. 156
— his death the following year, i. 159
Hutchinson, Mr., 'the oracle,' i. 139

'ILLUSTRATED NEWS,' picture of the troubles consequent on the break of gauge laid before the parliamentary committee, ii. 27
Institute of Civil Engineers, Robert Stephenson elected to the presidential chair of the, ii. 231
Ireland, state of, between Dublin and Cork in 1823, i. 62
— famine in, in 1846, ii. 131–134
— Lord George Bentinck's proposal to ameliorate it, ii. 131
Iron, cast, results of experiments on different kinds and different mixtures of, ii. 126
Iron Bridges. *See* Bridges, iron
— cast iron for arches, ii. 36
— wrought iron, its properties as distinguished from those of cast iron, ii. 37
— power of wrought iron to resist tension as compared with that of cast iron, ii. 87
Iron girders, different varieties of, ii. 46, 47
Iron ships, instance of the strength of the hulls of, ii. 298

JACKSON, Isaac, the clockmaker, Robert Stephenson's visit to, ii. 241
Jackson, William, one of the contractors for building the Victoria viaduct over the St. Lawrence, ii. 187, 209

JAM

James, Messrs., their locomotive boiler tubes, i. 51, 52

Jamieson, Professor, takes Robert Stephenson with him on a geological tour, i. 56

Jesmond Vale, i. 13

KAFFRE - AZZAYAT tubular bridge over the Nile, ii. 174
— completed by Mr. Rouse, ii. 244

Keefer, Mr. Thomas C., his survey for a bridge across the St. Lawrence, ii. 199

Kell, Mr., his description of Robert Stephenson at sea, ii. 172
— joins Robert Stephenson on a yachting excursion, ii. 241

Keogh, Judge, ii. 242

Kingstown and Dalkey Railway, trial of the atmospheric system on the, i. 310, 338

Killingworth, George Stephenson brakesman at, i. 10, 13
— appointed engine-wright at, i. 8
— colliery explosion in, i. 52

Killingworth Railway, trial of the 'Rocket' on the, i. 140

Kilsby tunnel, i. 200
— cost and labour expended upon it, i. 201

LA GUAYRA, appearance of the town of, i. 77
— Robert Stephenson's report on the construction of a breakwater before the harbour of, i. 78
— — his advice as to the construction of a pier, i. 79
— — and of a railway between La Guayra and Caraccas, i. 80, 81

Laing, Mr., consulted by Lord George Bentinck as to railway works in Ireland, ii. 132

Lardner, Dr., his attack upon the Stephensons in the 'Edinburgh Review,' i. 174, note; 179
— refutation of his accusations by Mr. Charles Lawrence, i. 179
— his 'simple rules' on railways, i. 288
— before the committees of the House, i. 288

Lea river, iron bridge over the, at Tottenham, ii. 48

LIV

Leake, General, his death, ii. 252

Lean, Mr. Charles, i. 205

Lecount, Lieut., his comparison of the London and Birmingham Railway with the Great Pyramid, i. 209

Lee, Mr. F. R., accompanies Robert Stephenson to Egypt, ii. 174

Leeds, locomotive coal carriage to, in 1758, i. 265

Leeds and Bradford Railway, opening of the, ii. 129

Leicester and Swannington Railway, i. 155
— commencement of the line, i. 164
— its advantages to the town of Leicester, i. 165

Leopold I. of Belgium and the two Stephensons, i. 220
— confers the Order of Leopold on George and Robert Stephenson, i. 220, 221
— creates Robert Stephenson a Knight of the Order of Leopold, 253

Leslie, Professor, his testimonial to Robert Stephenson, i. 57, 59

Leviathan ship, Mr. Brunel's, ii. 244

Liddell family, their seat of Ravensworth Castle, i. 14

Liddell, the Hon. H. T., M.P., his statement respecting George Stephenson, i. 257, 258

Lime-quarries at Stanhope, i. 215

Liverpool, George Stephenson's account of, i. 75
— its trade with Ireland by direct steamers, ii. 74
— Robert Stephenson as to the water supply of, ii. 139

Liverpool and Manchester Railway, failure of the Act for making the, i. 95
— George Stephenson superintendent of the line, i. 111, 112
— cause of the defeat of its bill in 1825, i. 111
— passing of the bill, i. 112
— the line at Chat Moss, i. 112
— magnitude of the excavations for the, i. 115
— the question of locomotives *versus* stationary engines under consideration, i. 117 *et seq.*
— report of Messrs. Walker and Rastrick, i. 122
— premium offered for the best locomotive, i. 124

LIV

Liver. and Manch. Railway—*cont.*
— George Stephenson's account of
Mr. Huskisson's inspection of the
line, i. 156
— opening of the line, i. 157
— Mr. Huskisson killed, i. 159
— work of constructing the line compared with that of the London and
Birmingham, i. 187
— Messrs. Walker and Rastrick's
report on the comparative merits
of the locomotive and stationary
engine system, i. 294
— the locomotive adopted through
the persistency of George Stephenson, i. 295
Locke, William, George Stephenson's
letter to, i. 59
Locke, Mr. Joseph, publishes, in conjunction with Robert Stephenson, 'Observations on the Comparative Merits of Locomotive and
Fixed Engines,' i. 151, 295
— his kindness to George Stephenson, i. 152
— constructs the Great Junction
Railway, i. 239
— his opposition to the atmospheric
system of propulsion, i. 332
— his dislike of iron bridges, ii. 52
— his evidence on iron railway structures, ii. 58
— his assistance in raising the tubes
of the Britannia bridge, ii. 98
— his death, ii. 265
Locomotives, Trevithick's, i. 24
— Steele's, i. 24, 25
— George Stephenson's first, i.
32
— trial of locomotives on the Stockton and Darlington line, i. 111
— state of the locomotive in 1828,
i. 116
— question of locomotives *versus* stationary engines, i. 117, 118
— Robert Stephenson's efforts at
improvement, i. 114, 117, 118
— the multitubular boiler and steamblast of the 'Rocket,' i. 118, 119
— failure of the bent boiler-tubes,
i. 120
— Messrs. Walker and Rastrick's
reports on the advantages and
disadvantages of the locomotive
system, i. 122
— what the locomotive of 1829 could
accomplish, i. 122

LON

Locomotives—*continued*
— the adoption of stationary engines
in preference to locomotives recommended, i. 123
— premium offered for the best locomotive, i. 124
— 'battle of the locomotive,' i. 138
— construction of the 'Rocket,' i.
139
— its trial on the Killingworth Railway, i. 140
— comparison between the evaporative capability of the 'Rocket'
and Stephenson's patent engines
of 1849, i. 141
— arrival of the 'Rocket' at Liverpool, i. 141
— the contest at Rainhill, i. 141
— the four engines engaged for the
contest, i. 142, 143
— triumph of the 'Rocket,' i. 143,
144
— history of the 'blast-pipe,' i. 144
— Mr. Hedley Wylam's method of
letting off the waste steam, i. 144
— George Stephenson's mode in the
'Puffing Billy,' i. 145
— Messrs. Walker and Rastrick's
report on stationary engines as
compared with locomotives, i. 150
— Robert Stephenson and Joseph
Locke's answer, i. 151
— prices of engines in 1830, i. 154
— Robert Stephenson's opposition
to the use of locomotives in towns,
i. 229
— Blenkinsop's patent locomotives
for coal carriage to Leeds, i. 264
— the locomotive as compared with
the stationary engine system, i. 294.
See Railways
— their advantages over stationary
engines as at present acknowledged,
i. 360
Logan, Mr. W. E., his description of
the phenomena of the packing and
piling of the ice of the river St.
Lawrence, ii. 196
London and Birmingham Railway,
original project for constructing
the, i. 166
— George and Robert Stephenson
employed to make the surveys and
plans, i. 166
— the agreement signed by the Company and the two Stephensons,
i. 167, *note*

LON

Lond. and Birm. Railway—*continued*
— the various surveys, i. 168, 169
— opposition to the line, i. 169
— 'Investigator's' pamphlet, i. 170
— the London and Birmingham Bill contest, i. 172
— the bill rejected by the Lords, i. 178
— public meeting at the Thatched House Tavern, i. 179
— passing of the bill, i. 180, 184, 185
— appointment of Robert Stephenson as sole engineer in chief, i. 181
— magnitude of the undertaking, i. 187
— Robert Stephenson's plans of every part of the entire line, i. 188, 189
— the staff of assistant and sub-assistant engineers, i. 190
— the drawing office at the Eyre Arms Hotel, i. 191
— list of the contracts and contractors, i. 192
— cost of the Primrose Hill tunnels, i. 194
— and of the Blisworth cutting, i. 194
— mishaps at the Wolverton embankment, i. 195
— vexatious opposition to the Company, i. 195
— defeat of the Grand Junction Canal Company, i. 200
— cost and labour expended upon the Kilsby tunnel, i. 201
— opposition of Lord Southampton, i. 206
— the stationary engines and ropes at Euston Square and Camden Town, i. 208
— the London and Birmingham Railway compared with the Great Pyramid, i. 209
— opening of the line, i. 210
— Robert Stephenson's anger with one of the directors, i. 211
— the electric telegraph used on the London and Birmingham line, i. 235
— proposal with respect to the London terminus in 1833, ii. 2
— defeated by the Great Western line, ii. 11
— the bowstring girder bridge over the Grand Junction Canal at Weedon, ii. 47

MAY

Lond. and Birm. Railway—*continued*
— report of Robert Stephenson on the question of gauge, ii. 272
London and Blackwall Railway, Robert Stephenson appointed engineer of the, i. 228
London and Brighton, various proposed lines of railway between, i. 221, 222
— Captain Alderson's report, i. 225
London and Colchester Railway, Mr. Braithwaite appointed engineer of the, i. 238
London and Southampton Railway, Mr. Francis Giles lays down the, i. 238
London Spring Water Company Bill, ii. 146
Longridge, Mr. Michael, Robert Stephenson's letters to, i. 57, 61, 62, 73, 75, 94, 101, 121; ii. 155
— joins the firm of Robert Stephenson & Co. i. 65
— takes the management during the absence of Robert Stephenson in South America, i. 100
Losh, Mr., his kindness to George Stephenson, i. 32, 33; ii. 181
— his rupture with George Stephenson, i. 65

MACLEAR, Sir Thomas, at the Royal Society Club, ii. 252
Magnay, Messrs., Robert Stephenson's paper-drying machine for, i. 71
Manby, Mr. Charles, in France, i. 26
— accompanies Robert Stephenson on a visit to the North, ii. 281
Manchester and Leeds Railway, George Stephenson appointed engineer of the, i. 239
Maps, English Government, ii. 147
Mariquita, the mines and mineralogical wealth of, i. 82
— Robert Stephenson's visit to, i. 82
— effects of earthquakes, stagnation of trade, and disturbed politics on the town, i. 83
— road from, to Honda, i. 83
— the two wet seasons at, i. 96
Marochetti, Baron, his statue of Robert Stephenson, ii. 267
May, Mr. Charles, his evidence on iron railway structures, ii. 58

MEC

Mechanical skill, Robert Stephenson's admiration of, ii. 232, 233

Mediterranean, its level compared with that of the Red Sea, i. 240; ii. 149, 151

Memorial, the Stephenson, at Willington Quay, i. 10

Menai Straits, width of the waterway in the strait, ii. 74
— interruptions of the traffic between England and Ireland at the, ii. 75
— various schemes for a roadway across the, ii. 75
— Government survey under Mr. Rennie, ii. 75
— his four designs for a bridge over the straits, ii. 75
— Mr. Telford's survey of the roads and straits, ii. 75, 76
— — his bridge over the, ii. 77
— rejection of Robert Stephenson's proposal for a bridge of two arches over the straits, ii. 81
— the bill for constructing the Britannia bridge passed, ii. 84
— laying of the first stone, ii. 89
— completion and opening of the bridge, ii. 100

Metropolitan Burials Bill, discussion on the, ii. 146

Montreal, grand banquet given to Robert Stephenson at, ii. 181

Montreal and Kingston Railway, construction of the, ii. 192

Moorsom, Captain (afterwards Admiral), his confidence in Robert Stephenson, i. 203, 204

Morton, Mr. A. C., his soundings of the river St. Lawrence, ii. 198

Minories, iron bridge over the, ii. 48

Multitubular locomotive boilers of Mr. Henry Booth, i. 52
— in locomotive engines, i. 118, 119
— Tredgold's account quoted, i. 128, note
— the 'Rocket' the first locomotive with a multitubular boiler, i. 144

NASMYTH, Mr., his steam piledriver, ii. 125

Navvies, the army of, employed on the London and Birmingham Railway, i. 201, 205

Neath Abbey works, i. 70

Nene Valley Drainage and Navigation Improvement Committee, Robert

NOR

Stephenson's connection with the, ii. 175, 176

Newcastle-on-Tyne, the Stephensons of, i. 1
— Hawthorn's locomotive factory at, i. 6
— the road from Newcastle to Killingworth, i. 13
— establishment of the firm of Robert Stephenson & Co. at, i. 65
— early difficulties of the works, i. 76
— the Literary and Philosophical Society of, i. 97, note
— the grand central railway station opened by the Queen, ii. 139
— account of the High-Level bridge at, ii. 113
— Robert Stephenson's donation to the Philosophical and Literary Institution, ii. 235
— and to the Wellington Memorial Schools, ii. 235
— divine service at Newcastle on the day of the funeral of Robert Stephenson, ii. 265

Newcastle and Berwick Railway, ii. 116, 129
— promoters take up the High-Level bridge, ii. 116

Newcastle and Carlisle Railway, i. 112, 117

Newcastle and Darlington Railway bill passed, i. 253
— opening of the line, i. 152, 257

New York, Robert Stephenson's opinion of the people of, i. 109

Niagara, Falls of, i. 109

Nicholson, Mr., his opposition to the atmospheric system of propulsion, i. 332

Nile, Robert Stephenson's iron bridge over the Damietta branch of the, ii. 69
— the great railway pontoon on the, ii. 174
— the Kaffre-Azzayat tubular bridge over the, ii. 174
— Robert Stephenson's description of the two tubular bridges over the, ii. 177
— the Kaffre-Azzayat viaduct, ii. 244

Nixon, Charles, of Walbottle colliery, i. 59

Norfolk Estuary Company, their dispute with the Eau Brink Commissioners, ii. 137

NOR

Norfolk Estuary Company—*cont.*
— their works, ii. 137
North Midland Railway, George Stephenson appointed engineer of part of the, i. 239
Norway, visit to, ii. 130
— railway from Christiania to the Myren Lake constructed, ii. 130
Nubar Bey visits London, ii. 174

OGWEN RIVER, viaduct over the, ii. 79
Ordnance maps, Robert Stephenson's speech on the, ii. 147
Orme's Head, Great and Little, ii. 78
'Our Club,' ii. 230
Oxford and Wolverhampton Railway bills, ii. 10

PAINE, Tom, his cast-iron bridge, ii. 34, 35, 233
— Robert Stephenson's restoration of this bridge, ii. 70, 71
— description of it, ii. 234
Palmerston, Viscount, his speech on the Suez Canal scheme, ii. 148
Panama, Isthmus of, Robert Stephenson's desire to visit the, i. 102
Papin, Denys, his notion of the atmospheric system of propulsion, i. 299
Paris and St. Germain Railway, trial of the atmospheric system on the, i. 351
Parker, Mr. Charles, i. 206
Pease, Mr. Edward, George Stephenson's first interview with, i. 53
— joins the firm of Robert Stephenson & Co., i. 65
Pease, Mr. Joseph, his recollections of George and Robert Stephenson, i. 54
Peel, Sir Robert, at the opening of the Liverpool and Manchester Railway, i. 157
— his support of the atmospheric plan, i. 310
Penmaen Bach and Penmaen Mawr, railway through the, ii. 79
Percy, Dr., Robert Stephenson's visits to, ii. 250
Perkins's engine, George Stephenson's criticism on, referred to, i. 62

RAI

Peto, Samuel Morton, one of the contractors for building the Victoria viaduct over the St. Lawrence, ii. 187, 209
Phillips, Richard, teaches mineralogical chemistry to Robert Stephenson, i. 72
Phipps, Mr. G. H., i. 139, 148
— letter from George Stephenson to, concerning the blast-pipe, i. 148
Pier at La Guayra, Robert Stephenson advises the construction of a, i. 79
Piers of the Victoria bridge over the St. Lawrence, ii. 217, 221
Pile-driver, Nasmyth's steam, ii. 125
Pim, Mr. James, his advocacy of the atmospheric system, i. 309
— obtains a trial for it on the Kingstown and Dalkey Railway, i. 310
Pinkus, Mr. Henry, his patent atmospheric tubes, i. 301
Plymouth, Mr. J. M. Rendel's chain ferry-boat at, ii. 174
Pneumatic system. *See* Atmospheric system of railway propulsion
Pontop and South Shields Railway bill passed, i. 250, 252
— its subsequent success, i. 253
Portland, the terminus of the Grand Trunk Railway of Canada at the harbour of, ii. 192, 193
— inconveniences which may arise at some future time, ii. 193
Powell, Rev. Baden, his death, ii. 252
Preston, Mr., surveyor to the Alexandria and Cairo Railway, ii. 179, *note*
Primrose Hill tunnel, difficulties of making the, i. 193
— contract price and cost of, i. 194
'Prince of Wales' iron steam-vessel, accident to the, ii. 298, *note*
Pringle, Mr., in Egypt, ii. 179, *note*
'Puffing Billy,' the, i. 145, 146
Pym, Captain, his paper on the Suez Canal question, ii. 250

QUEBEC and Richmond Railway, formation of the, ii. 192

RAILWAY between Caraccas and La Guayra, Robert Stephenson's report on the proposed, i. 80, 81

RAI

Railway contractors in 1833, i. 189, 193

Railway, the Commercial Road, i. 228

Railway Society, the, i. 240

Railways *versus* Coaches, i. 170, 171

— in Belgium, i. 220

— railway enterprise in England in 1836, 1837, i. 221

— establishment of the railway system, i. 238

— railways undertaken in various directions, i. 238

— the engineers engaged on the great lines, i. 238

— the Great Western and Mr. Brunel, i. 238

— the London and Southampton and Mr. Francis Giles, i. 238

— the London and Colchester and Mr. Braithwaite, i. 238

— the various lines in the hands of George Stephenson, i. 239

— the Pontop and South Shields line, i. 252, 253

— the Newcastle and Darlington line, i. 251, 252, 256

— history of railway progress and railway legislation, i. 263

— the first Act of Parliament authorising the construction of a railway, i. 264

— the Surrey Iron Railway Company, i. 264

— table of Railway Acts passed from 1801 to 1840, i. 264, 265

— the railway mania of 1844, i. 266

— amount of capital assigned by Acts of Parliament in 1844, 1845, and 1846 to railway enterprise, i. 267

— corrupt legislation, i. 268

— the peer, his park, and compensation, i. 269

— resolution of Parliament, i. 273

— the three periods of railway mania, i. 272

— the bubble companies of 1836, i. 274

— the ten-and-sixpenny capitalists, i. 276

— parliamentary votes sold for gold by British senators, i. 277, 278

— extortions of all classes of society, i. 278

— national benefits of railways, i. 279

— amicable co-operation of members of Parliament in 1845, i. 280

RAI

Railways—*continued*

— Robert Stephenson's presidential address, i. 281

— great number of Acts regarding railways, i. 281

— vast expenses of obtaining permission from Parliament to construct railways, i. 283

—Robert Stephenson's remedy, i. 284

— parliamentary barristers and their enormous fees, i. 285, 286

— venality of the testimony of some of the engineers in the committee rooms of the House, i. 287

— Dr. Lardner's 'simple rules' on railways, i. 288

— railway development in Great Britain as compared with that of other countries, i. 289

— Mr. Bridge Adams's work, i. 290

— ruinous contests of rival lines, i. 290

— proposal for railway farmers, i. 291

— history of the atmospheric system of railway traction, i. 292

— Robert Stephenson's constant opposition to it, i. 293

— three modes of locomotion upon railways, i. 293

— constant rivalry between locomotive and stationary steam power, i. 293

— Messrs. Walker and Rastrick's report on locomotives as compared with stationary engines, i. 294

— Robert Stephenson and Joseph Locke's tract on the locomotive system, i. 295

— first trial of the atmospheric system on the Birmingham, Bristol, and Thames Junction Railway, i. 303

— Robert Stephenson's report and comparison between the atmospheric and stationary engine and locomotive systems, i. 316 *et seq.*

— interest of the public in the atmospheric question, i. 329

— the railway mania of 1845, i. 330

— results of the various trials made of the atmospheric system, i. 356

— the battle of the gauges, ii. 1

— Brunel's theory of railway districts, ii. 8

— the first actual 'break of gauge,' ii. 9

RAI

Railways—*continued*
— the Oxford and Wolverhampton rival schemes, ii. 10
— failure of Brunel's theory of railway districts, ii. 11
— establishment of the Railway Clearing House, ii. 16
— table of the lines completed, in progress, and projected, in 1845, ii. 24
— history of the Chester and Holyhead Railway, and of the designs for the Britannia and Conway tubular bridges, ii. 285
— fatal accident to the bridge over the Dee, ii. 48
— engineering interest attached to the Chester and Holyhead Railway, ii. 78
— course of this line, ii. 78, 79
— opening of the Newcastle and Darlington Railway, ii. 115
— and of the Newcastle and Berwick and Trent Valley Railways, ii. 129
— construction of the Christiania and Myren Lake Railway, ii. 130
— the Alexandria and Cairo Railway, ii. 173
— journey on a North American railway, ii. 180
— a Bogie engine, ii. 180
— Robert Stephenson's speech on railway legislation in England, ii. 182
— the first railway efforts in Canada, ii. 182
— the Atlantic and St. Lawrence Railway, ii. 183
— other lines in Canada, ii. 183
— formation of the Grand Trunk Railway of Canada, ii. 193
— reports of Brunel and Stephenson on the question of gauge, ii. 269, 272
Rastrick, Mr., his cast-iron girders at the British Museum, ii. 44
— his evidence on iron railway structures, ii. 58
Ravensworth, Lord, his kindness to George Stephenson, ii. 143
Ravensworth Castle, i. 14
Red Sea, its level compared with that of the Mediterranean, i. 240; ii. 149, 151
Rendel, Mr., his mode of trussing for the prevention of oscillation of

ROS

the platform of suspension bridges, ii. 292
Rendel, Mr. J. M., his chain ferryboat at Plymouth, ii. 174
Rennie, Sir John, his proposed railway between London and Brighton, i. 222
— appointed to survey the Menai Straits, ii. 75
— his four designs for a bridge, ii. 75
— appointed one of the engineers-inchief to the Norfolk Estuary Company, ii. 139
Richardson, Mr. Thomas, joins the firm of Robert Stephenson & Co., i. 65, 67
— one of the projectors of the Colombian Mining Association, i. 67
— his interest in Robert Stephenson's welfare, i. 134
— Stephenson's letters to him, i. 135, 137, 151
Rivière du Loup, terminus of the Grand Trunk Railway of Canada at, ii. 193
Robertson, Mr., his opinion of the cause of the Dee bridge accident, ii. 52
Robinson, Rev. Dr., his support of the principle of atmospheric propulsion, i. 332
Robson, Captain, his intimacy with George Stephenson, i. 27
— his firm at Killingworth, i. 27
— his account of George Stephenson's safety-lamp, i. 27
'Rocket,' the, i. 118, 127–130, *note*
— construction of the engine, i. 138
— failure of the first boiler, i. 139, 140
— its trial on the Killingworth Railway, i. 140
— shipped to Liverpool, i. 141
— the contest at Rainhill, i. 141
— particulars of the 'Rocket,' i. 142, 143
— the 'Rocket' proclaimed the winner, i. 144
— the blast-pipe of the 'Rocket,' i. 144
Roebuck, Mr., his motion respecting the Suez Canal, ii. 150
Roman bridges, ii. 32
Ross, Mr. Alexander, co-engineer of the Victoria (St. Lawrence) bridge, ii. 179, 180, 187

ROS

Ross, Mr. A.—*continued*
— his engineering abilities, ii. 199
— examines the various points of crossing proposed for the Victoria bridge, ii. 200
— visits England and takes counsel with Robert Stephenson, ii. 200
— returns to Canada and prepares a report to the Railway Commissioners, ii. 201
— constructs the Victoria bridge, ii. 244
Roullin, Dr., visits Robert Stephenson at Santa Ana, i. 92
Rouse, Mr. Henry J., in Egypt, ii. 179, *note*
— and the Kaffre-Azzayat viaduct over the Nile, ii. 244
Royal Society, Robert Stephenson elected a fellow of, ii. 137
Royal Society Club, Robert Stephenson at the dinners of the, ii. 230
— Robert Stephenson's last visit to the, ii. 252
Rushton, Mr., in Egypt, ii. 179, *note*
Russell, Lord John, Robert Stephenson's remark on, ii. 144
Rutter, Tommy, his school at Long Benton, i. 19, 20
— Robert Stephenson a pupil there, i. 20
Ryle, Robert Stephenson's visits to, i. 19

SAFETY-LAMP, Captain Robson's account of George Stephenson's, i. 27
— the safety-lamps of Dr. Clanny, Sir Humphry Davy, and George Stephenson, i. 37
Said Pacha, viceroy of Egypt, in Abyssinia, ii. 247
— his letter to Robert Stephenson, ii. 261
Salta Falls, i. 110
Saltash iron bridge, ii. 67
Samuda, Messrs., their advocacy of the atmospheric system of propulsion, i. 302
— with Mr. Clegg, lay down a short line of atmospheric tubes, i. 303
— trial of their system, i. 303
— their calculation of the expense of atmospheric tubes and engines per mile, i. 324

STA

Samuda, Messrs.—*continued*
— Mr. Jacob Samuda's evidence before the House of Commons Committee, i. 331
Sandars, Mr. Joseph, i. 164
Sanders, Mr., of Liverpool, i. 75
Sanderson, Miss Fanny, i. 132
— her marriage to Robert Stephenson, i. 137
— her personal appearance, i. 137
— her death, i. 255
Sanderson, Mr. John, and the crackbrained projectors, ii. 169
— his death, ii. 186
Santa Ana, Robert Stephenson at the mines of, i. 84
— his description of the scenery of, i. 85
— the Cornish miners at, i. 87, 88
— Stephenson's cottage at, i. 91
— his departure from, i. 100
Santa Fé de Bogota, state of the road to, in 1825, i. 81, 82
Serpentine, Robert Stephenson's speech on the state of the, ii. 152
Severn, iron bridges over the, ii. 33, 34
Sibthorp, Colonel, his speech and motion against the proposal for the Great Exhibition of 1851, ii. 146
Smeaton, his atmospheric engine at Long Benton, i. 12
— his consultation with Robert Stephenson about the Hull Docks, i. 254
Smith, Lieut.-Col. Sir F., his favourable report on the atmospheric system, i. 310
— on the Royal Gauge Commission, ii. 11
Smyth, Professor P., his observations at Teneriffe, ii. 172
Snibston collieries, purchase of the, i. 164
Sopwith, Mr., accompanies Robert Stephenson to Egypt, ii. 174
— his account of the steam ferry on the Nile, ii. 174
Southampton, Lord, his opposition to the London and Birmingham Railway, i. 206
South Devon Railway, application of the atmospheric system to the, i. 346
Southwark iron bridge, ii. 36
Stanhope and Tyne Railway, i. 215

STA

Stanhope and Tyne Railway—*cont.*
— Robert Stephenson appointed engineer, i. 217
— crisis in the affairs of the Company, i. 245
— plan for meeting the difficulty, i. 248
— dissolution of the Company, i. 248
— reformed under the name of the Pontop and South Shields Railway Company, i. 251
Stanton, Mr. Philip, i. 248
Stanton, Mr. J. H., in Egypt, ii. 179, *note*
Steam coaches, i. 115
Steele, John, .his intimacy with George Stephenson, i. 23
— notice of his career, i. 23-25
— his death, i. 26
Stephensons, the, of Newcastle, i. 1
Stephenson, 'Old Robert,' i. 2
— his marriage, i. 2
— his children, i. 2
— struck blind, i. 16
Stephenson, George, his parentage, i. 2
— his early life, i. 3
— learns the art of shoe-cobbling, i. 3
— becomes brakesman at the Dolly Pit, i. 3
— his acquaintance with Elizabeth Hindmarsh, i. 3
— woos Ann Henderson in vain, i. 3, 4
— marries Fanny Henderson, i. 5
— goes to Willington Quay, i. 6
— first sets up housekeeping, i. 6
— applies himself to the work of self-education, i. 7
— birth of his son Robert, i. 7
— the christening festivities, i. 7
— his wife's delicate health, i. 8
— beginning of his intimacy with Robert Hawthorn, i. 9
— learns clock-mending and clock-cleaning, i. 9
— removes to Killingworth, i. 10
— his cottage there, i. 13, 14
— death of his wife, i. 15
— goes to Montrose, i. 15
— returns to the West Moor, i. 16
— his dislike of 'artificials,' i. 17
— his early friends, i. 23
— Captain Robson's account of Stephenson's safety-lamp, i. 27

STE

Stephenson, George—*continued*
— Stephenson appointed enginewright of the Killingworth colliery, i. 30
— his first locomotive engine, i. 32
— his engagement at the Walker iron works, i. 33
— sends his son to Mr. Bruce's school, i. 34
— invention of his safety-lamp, i. 37
— contest between Sir H. Davy and Stephenson, i. 37
— story of his foolhardy trial of his lamp, i. 38
— testimonial and public dinner to him for his invention, i. 39
— his speech in returning thanks, i. 40
— his attempts to learn grammar, i. 41, *note*
— his notion of the amount of education necessary for his son, i. 43
— his sun-dial at the West Moor cottage, i. 44
— lays down and completes the Hetton Colliery Railway, i. 49
— his increasing prosperity, i. 49
— marries Elizabeth Hindmarsh, i. 50
— builds the Friar's Goose pumping engine, i. 51
— embarks in a colliery speculation, i. 51
— his first visit to Mr. Edward Pease, i. 53
— engaged as engineer-in-chief of the Stockton and Darlington Railway, i. 53
— puts his son's name on the map of the Stockton and Darlington line, i. 55
— sends Robert to Edinburgh University, i. 55
— his letter to his friend William Locke, i. 59
— goes to Ireland, i. 61
— his criticism on Perkins's engine, i. 62
— his rupture with Mr. Losh, i. 64
— succeeds in establishing the locomotive engine factory of Robert Stephenson & Co., i. 65
— his engine for drawing coals at Swansea, i. 70
— gives his reluctant consent to Robert's journey to Colombia, i. 70

STE

Stephenson, George—*continued*
— takes leave of his son at Liverpool, i. 75
— his entertaining letter to Mr. Longridge, i. 75
— employed as engineer-in-chief to the Liverpool and Manchester Railway, i. 111, 112
— his difficulties at Chat Moss, i. 112
— return of his son to England, i. 113
— the multitubular boiler, i. 119–126
— his letter to Mr. Phipps on his exhausting pipe, i. 148
— Joseph Locke's kindness to George Stephenson, i. 152
— George Stephenson's account of Mr. Huskisson's inspection of the Liverpool and Manchester Railway, i. 156
— opening of the line, i. 157
— purchase of the Snibston collieries, i. 164
— employed with his son to make the surveys and plans for the London and Birmingham Railway, i. 166
— the agreement entered into, i. 167, *note*
— accompanies his son to Brussels, i. 220
— his reception by the Belgians, i. 220
— decorated with the Order of Leopold, i. 220
— the various railways on his hands in 1839, i. 239
— his collieries at Clay Cross, i. 240
— present at the dinner given to his son at the Albion, i. 242
— and at the banquet on the opening of the Newcastle and Darlington Railway, i. 258
— erroneous statement of Mr. Liddell respecting him, i. 258
— instance of his quick thought and care for his own interests, i. 260
— his trial of both the locomotive and stationary steam-power systems, i. 294
— his persistent defence of the locomotive system, i. 295
— his iron railway bridges on the Manchester and Liverpool Railway, ii. 45

STE

Stephenson, George—*continued*
— his third marriage and death, ii. 135
— affectionate relations between George and Robert Stephenson, ii. 136
— the father's wealth, ii. 136
— his political opinions and sympathies, ii. 142
— gauge of his first public railway, ii. 3
— his chest of drawers in the cottage at Long Benton, ii. 239
STEPHENSON, ROBERT, his birth, i. 7
— his earliest recollections of his father, i. 8
— memorial of his father's tenderness, i. 9
— death of his mother, i. 15
— his aunt Eleanor housekeeper at West Moor, i. 17
— his visits to his relations scattered about the country, i. 18
— sent to Tommy Rutter's school at Long Benton, i. 20
— goes gleaning with his aunt Eleanor, i. 21
— his first commissions, i. 29
— sent to Mr. Bruce's school, in Newcastle, i. 34
— his personal appearance at this time, i. 34
— his delicate health, i. 35
— his donkey and blackbird, i. 36
— his friend John Tate, i. 36
— his work and amusement, i. 42
— throwing the hammer, i. 42
— his father's notions as to his education, i. 43
— learns French, i. 43
— his drawing of a sun-dial, i. 44
— story of his electrical kite, i. 44
— leaves school and is apprenticed to Mr. Nicholas Wood, i. 46
— his hard work and careful economy,
— his meals at the 'Three Tuns,' i. 48
— his circumferentor, i. 48
— learns music, i. 50
— in the Killingworth mine during an explosion, i. 52
— accompanies and assists his father in surveying the Stockton and Darlington Railway, i. 53, 54
— his visit to London, i. 53, 54
— his delicate health, i. 54
— his name put to the map of the Stockton and Darlington line, i. 55

STE

Stephenson, Robert—*continued*

— his residence in the university of Edinburgh, i. 55
— accompanies Professor Jamieson on a geological excursion, i. 56
— Professor Leslie's testimonial, i. 57, 59
— his letters to Mr. Michael Longridge, i. 57, 61, 62, 73, 75, 94, 101, 121
— commencement of his friendship with George Parker Bidder, i. 60
— accompanies his father to Ireland, i. 61
— becomes a partner and prominent engineer in the firm of Robert Stephenson & Co., i. 65
— asked to accompany the expedition of the Colombian Mining Association, i. 68
— takes a professional trip into Cornwall and other places, i. 69
— his letters to his father, i. 70, 74
— obtains his father's consent for a voyage to Colombia, i. 70
— his preparations for the voyage, i. 71
— his coining press and paper-drying machine, i. 71
— takes lessons in mineralogical chemistry, i. 72
— takes the coach to Liverpool, i. 74
— takes leave of his father there, i. 75
— sets apart 300*l*. per annum for his father, i. 76
— sails for South America, i. 76
— lands at La Guayra, i. 78
— his report against the construction of a breakwater before the harbour of La Guayra, i. 78
— advises the building of a pier in the harbour, i. 79
— difficulties of a railway between Caraccas and La Guayra, i. 80, 81
— goes to Santa Fé de Bogota, i. 81, 82
— examines the mines at Mariquita, i. 82
— compelled to leave the heavier portion of his machinery behind, i. 84
— commences working for ore, i. 84
— his letters to his stepmother, i. 85, 93
— saves a drowning man, i. 86
— his difficulties with the Cornish miners at Santa Ana, i. 87, 88

STE

Stephenson, Robert—*continued*

— their threats, i. 89
— his treatment of them, i. 90
— builds a cottage at Santa Ana, i. 91
— visited by M. Boussingault and Dr. Roullin, i. 92
— his correspondents in England, i. 93
— convinced of the visionary nature of the enterprise at which he was at the head, i. 93, 94
— his inclination and duty, i. 95
— leaves Santa Ana, i. 100
— wishes to visit Panama, i. 102
— goes up to Carthagena, i. 105
— sails for New York, i. 105
— becalmed, i. 106
— shipwrecked crews and cannibalism, i. 106
— in a storm at sea, i. 107
— loses his entomological curiosities and his money, i. 107
— arrives in New York, i. 108
— makes a pedestrian excursion from New York to Montreal, i. 108
— his opinion of Canada, i. 110
— sails for England, i. 111
— position of his father at this time, i. 111
— his report to the Colombian directors in London, i. 114
— his trip to Brussels, i. 114
— his efforts at improving the locomotive, i. 114, 117, 118
— his residence in Newcastle, i. 116
— his modes of increasing the heating surface of his boilers, i. 118
— trial of his new boiler made to burn coke, i. 121
— premium of 500*l*. offered by the Liverpool and Manchester Railway for the best locomotive, i. 124
— Robert Stephenson receives a sketch of the multitubular boiler from his father, i. 126
— account given by Robert Stephenson to Mr. Smiles, i. 126, *note*
— his locomotive the 'Rocket,' i. 118, 127–130, *note*
— business on his hands in 1828, 1829, i. 130
— offers to Miss Fanny Sanderson, i. 133
— his first house at Greenfield Place, Newcastle, i. 134
— his letter of defence to Mr. Richardson, i. 135

STE

Stephenson, Robert—*continued*
— his marriage, i. 137
— his wedding trip, i. 138
— failure of the first boiler of the 'Rocket,' i. 139, 140
— his discovery of the right method, i. 140
— the trial of the 'Rocket' on the Killingworth Railway, i. 140
— arrival of the 'Rocket' in Liverpool, i. 141
— the contest at Rainhill, i. 141
— the 'Rocket' proclaimed the winner, i. 144
— history of the blast-pipe, i. 144
— triumphant return to Newcastle, i. 150
— his 'Observations on the Comparative Merits of Locomotive and Fixed Engines,' i. 151
— his letters to Mr. Richardson, i. 151
— his numerous engagements, i. 152
— at the opening of the Liverpool and Manchester Railway, i. 158
— appointed engineer to the Warrington and Leicester and Swannington Railways, i. 164
— becomes a member of the Institute of Civil Engineers, i. 165
— employed, with his father, to make the surveys and plans for the London and Birmingham Railway, i. 166
— the agreement entered into, i. 167, note
— his three surveys, i. 168, 169
— opposition to the line, i. 169
— his cross-examination before the Lords' Committee, i. 173–176, note
— his exertions to get the bill through Parliament, i. 177
— his mortification at the rejection of the bill, i. 178
— kindness of the chairman of the committee, Lord Wharncliffe, i. 178
— passing of the London and Birmingham bill, i. 180
— appointed sole engineer-in-chief of the line, i. 181, 185
— leaves Newcastle-on-Tyne, and takes a house on Haverstock Hill, i. 181
— his evening parties at Greenfield Place, i. 182

STE

Stephenson, Robert—*continued*
— his pupils, i. 182, 183
— stupendous nature of the task undertaken by him in making the London and Birmingham Railway, i. 187
— health and habits of life, i. 187
— his mode of checking the work of the contractors, i. 189
— his staff of assistant and sub-assistant engineers, i. 190
— his difficulties with the Primrose Hill tunnel, i. 193
— accidents at the Wolverton embankment, i. 195
— his temporary bridge over the Grand Junction Canal, i. 198
— the Kilsby tunnel, i. 200
— his interview with Dr. Arnold at Rugby, i. 202
— difficulties with the Kilsby tunnel, i. 202 *et seq.*
— his suggestion of the advisability of carrying the line into London, i. 206
— snubbed for his proposal, i. 207
— completion of his work, i. 210
— opening of the London and Birmingham Railway, i. 210
— insult offered to his father, i. 210
— his anger, i. 211
— dinner and testimonial given to him at the Dun Cow, at Dunchurch, i. 211
— appointed consulting engineer to the London and Birmingham, i. 212
— his system of drawing used by Brunel for the Great Western, i. 213
— his occupations during the progress of the London and Birmingham line, i. 214
— his connection with the Stanhope and Tyne Railway, i. 215
— accompanies his father to Belgium, i. 220
— decorated with the Order of Leopold, i. 221
— takes offices in Duke Street and Great George Street, Westminster, i. 221
— projects, and becomes engineer of, a line from London to Brighton, i. 221 *et seq.*
— his and Gibbs's lines rejected by Capt. Alderson, i. 226

Stephenson, Robert—*continued*

— defends his views in a pamphlet, i. 226

— appointed engineer of the London and Blackwall Railway, i. 228

— his opposition to the use of locomotives in towns, i. 229

— his evidence on the subject before a House of Commons' committee, i. 229, *note*

— his life at Haverstock Hill, i. 231

— his political convictions, i. 232

— Mrs. Stephenson, i. 232, 233

— his Newcastle correspondence i. 233

— Mrs. Stephenson's accident to her kneecap, i. 234

— his most intimate friends, i. 235

— his adoption of the electric telegraph, i. 235

— assumes arms, i. 237

— visits the Continent, i. 239

— his intimacy with Mons. Talabot, i. 240

— their survey of the Isthmus of Suez, i. 240

— his immersion in business on his return home, i. 240 .

— contractors' testimonial and dinner at the Albion Tavern, i. 241, 242

— his. arbitration in disputes between the contractors and the companies, i. 241, 242

— his letters to Mr. Cook, i. 243, 245, 249

— his anxiety as to the affairs of the Stanhope and Tyne Railway, i. 245

— his astonishment at his responsibilities, i. 245, 246

— appointed engineer to the Newcastle and Darlington Railway, i. 253

— created by the King of the Belgians a Knight of the Order of Leopold, i. 253

— death of Mrs. Stephenson, i, 255

— his consultation about the Hull Docks, i. 254

— rejoicings on the opening of the Newcastle and Darlington line, i. 256

— bridge over the Wear, i. 257

— the public dinner and speeches at Newcastle, i. 257

— his Continental engagements, i. 259

Stephenson, Robert—*continued*

— removes to Cambridge Square, i. 259

— fire in his new residence, i. 259

— review of his position at forty years of age, i. 260

— his advocacy of the creation of a railway board of enquiry, i. 280

— his presidential address in 1856, i. 281

— his conscientious evidence before parliamentary committees, i. 287

— his opposition to the atmospheric system, i. 293, 332

— his part in a tract in favour of the locomotive and against the stationary engine system, i. 295

— his attention called to the atmospheric system in reference to the Chester and Holyhead Railway, i. 315

— his report to the Chester and Holyhead directors, i. 316

— his friendship for Brunel, ii. 2

— his character as a parliamentary witness, ii. 20

— his examination by the Gauge Commissioners, ii. 21

— his remarks on Brunel's loose box system, ii. 22

— his objections to the double gauge system, ii. 23, 29

— his history of the Britannia and Conway tubular bridges, ii. 285

— his small bridge on the Hertford and Ware Branch Railway, ii. 294

— services rendered him by Mr. William Fairbairn and Mr. Eaton Hodgkinson, ii. 299, 300, 303

— his great practice in iron bridges, ii. 30

— his article on iron bridges in the 'Encyclopædia Britannica,' ii. 31

— accident to his bridge over the Dee at Chester, ii. 48

— his description of the defects of compound girders, ii. 53

— his evidence before the Iron Railway Structure Commission, ii. 54

— his evidence before the royal commission on the subject of iron bridges, ii. 57

— his application of wrought-iron girders for bridge construction, ii. 62

— his iron bridge over the Aire, ii. 68

STE

Stephenson, Robert—*continued*
— his appointment as engineer of the Chester and Holyhead Railway, ii. 80
— his selection of the site of the Britannia bridge, ii. 80
— rejection of his proposal of a bridge of two arches over the Straits, ii. 81
— the first idea of the tubular combination, ii. 82
— passing of the bill authorising the construction of the Britannia bridge, ii. 84
— his experiments for the designs of the tube, ii. 85
— his report to the directors, ii. 86
— disheartening doubts as to efficiency of his proposed work, ii. 88
— laying of the first stone of the Britannia bridge, ii. 89
— his further experiments for the design of the tubes, ii. 89
— manufacture of the tubes, ii. 94
— floating and hoisting them up, ii. 96
— invitation to Mr. Brunel and Mr. Locke to aid in raising the tubes, ii. 98
— completion and opening of the bridge, ii. 100
— Mr. Stephenson's explanations of the peculiarities in the combinations, ii. 104
— description of his bridge at Conway, ii. 110
— account of his high-level bridge at Newcastle-on-Tyne, ii. 113
— his evidence before committee on the subject, ii. 116
— passing of the bill, ii. 117
— description of the bridge, ii. 117
— Mr. Stephenson's motives for the adoption of the bowstring girder, ii. 120, 121
— erection and completion of the bridge, ii. 127
— his work in 1845, 1846, ii. 129
— joins the Committee of the Great Exhibition of 1851, ii. 130
— visits Italy and Norway, ii. 130
— liberality of the Government of Norway, ii. 130, 131
— consulted by Lord G. Bentinck as to subsidising Irish railways, ii. 132

STE

Stephenson, Robert—*continued*
— elected member of Parliament for Whitby, ii. 134, 135
— elected to the council and vice-presidency of the Institute of Civil Engineers, ii. 135
— his narrow escape on the Chester and Holyhead Railway, ii. 135
— death of his father, ii. 135
— relations between father and son, ii. 136
— portrait of Robert in the possession of the Newcastle Literary and Philosophical Society, ii. 136
— elected a Fellow of the Royal Society, ii. 137
— his connection with the Nene Valley Drainage and Navigation Improvement Committee, ii. 137, 138
— appointed one of the engineers-in-chief of the Norfolk Estuary Company, ii. 138
— consulted by the Town Council of Liverpool as to the best means of supplying the city with water, ii. 137, 138
— his desire for rest, ii. 140
— his character as a politician and member of the House of Commons, ii. 142
— his Toryism, ii. 143
— his remarks on 'little Lord John,' ii. 144
— his opinions on popular education, ii. 144
— his belief in protective principles, ii. 145
— his speech in the House of Commons on the proposed site of the Great Exhibition of 1851, ii. 146
— and on the Victoria sewer, ii. 146
— his speech on the Ordnance maps, ii. 147
— and on the proposal for a Suez canal, ii. 149, 151
— his speech on the state of the Serpentine, ii. 152
— his popularity in the House of Commons, ii. 153
— his letter to Admiral Moorsom on the Crimean mismanagement, ii. 153, 154
— his reason for declining the invitation of the Newcastle Conservatives, ii. 154
— his dislike of party strife, ii. 155

STE

Stephenson, Robert—*continued*

— his part in the testimonial to Sir W. Hayter, ii. 155

— fullness of his prosperity, ii. 157, 166

— his personal appearance in 1851, ii. 158

— his visit to the impostor of Soho, ii. 158

— his house in Gloucester Square, ii. 159

— his Sunday lunches, ii. 159

— works of art contained in his house at the time of his death, ii. 160

— his philosophical apparatus and collections, ii. 160

— his demeanour in society, ii. 161

— and in his own profession, ii. 162

— his feudal sway in Great George Street, ii. 162

— his liberality to professional subordinates, ii. 162

— his earnings, ii. 164, 165

— his treatment of cabmen, ii. 165

— becomes a member of the Committee for the Exhibition Building of 1851, ii. 165

— intrusion of crack-brained projectors on his privacy, ii. 168

— his fondness for aquatic amusements, ii. 169, 170

— his yachts, ii. 170, 171

— Mr. Kell's account of him on board, ii. 172

— his visit to Egypt, ii. 173

— commissioned to construct the Alexandria and Cairo Railway, ii. 173

— his second visit to Egypt, ii. 174

— undertakes the construction of the Kaffre-Azzayat bridge, ii. 176

— his description of the bridge, ii. 177

— his engineering staff in Egypt, ii. 179, *note*

— entrusts the designing of the tubular bridges over the Nile to his cousin, Mr. G. R. Stephenson, ii. 179, *note*

— his trip to Canada, ii. 180

— engages to construct the Victoria viaduct over the St. Lawrence, ii. 180

— his journey on a North American railway, ii. 180

— grand banquet given to him at Montreal, ii. 181

— his speech on the occasion, ii. 181

STE

Stephenson, Robert—*continued*

— death of his brother-in-law, Mr. John Sanderson, ii. 186

— his connection with Mr. Alexander Ross, ii. 187

— arrangements with the engineers and contractors of the Victoria viaduct, ii. 187

— his frequent arbitrations between railways and contractors, ii. 188

— the details of the Victoria viaduct worked out in his office in Great George Street, ii. 189

— completion of the viaduct, ii. 189

— consulted by Mr. Ross as to the bridge over the St. Lawrence, ii. 200

— their report to the Railway Commissioners, Quebec, ii. 201

— Robert Stephenson's visit to Canada, and letter to the railway directors respecting the Victoria bridge, ii. 203

— his answer to charges of extravagance in his design, ii. 210

— sends out Mr. Stockman and Mr. S. P. Bidder to examine the progress of the work, ii. 226

— his fondness for the periodical dinners of learned societies, ii. 230, 231

— elected a member of the Royal Society Club, ii. 231

— occupies the Presidential chair of the Institution of Civil Engineers, ii. 231

— receives the honorary degree of D.C.L. of Oxford, ii. 231

— the dark side of his prosperity, ii. 232

— his admiration of mechanical skill, ii. 232, 233

— his speech at Sunderland, ii. 233

— his last work, ii. 234

— his donation to the Newcastle Philosophical and Literary Institution, ii. 235

— and to the Wellington Memorial Schools, ii. 235

— his periodical visits to the factory in Newcastle, ii. 235

— his considerate conduct to his humble relations, ii. 236

— his visits to the scenes of his youth, ii. 237

— goes on a yachting cruise with a few friends, ii. 241

STE

Stephenson, Robert—*continued*
— becomes a more general reader than before, ii. 243
— his increasing illness, ii. 243
— his inspiration and regulation of the labours of younger engineers, ii. 244
— his social engagements, ii. 244
— his trip to Egypt, ii. 244
— his letters from Alexandria, ii. 245, 246
— his delight in the climate of Egypt, ii. 246
— visits Algiers, ii. 247
— meets the Brunels at Cairo, ii. 248
— his last Christmas dinner there, ii. 249
— his last London season, ii. 249
— acts on the commission 'To Inquire into the Construction of Submarine Telegraph Cables,' ii. 250
— means suggested by him for extracting an enormous granite sarcophagus from a limestone cavity in Egypt, ii. 250
— present at the annual inspection of Greenwich Observatory, ii. 251
— his last visit to the Royal Society Club, ii. 251, 252
— his last will and testament, ii. 253
— his last voyage to Norway, ii. 254
— grand dinner given to him at Christiania, ii. 254
— his speech in returning thanks for the toast of his health, ii. 256
— his voyage home, ii. 258
— his illness on board, ii. 258
— lands at Lowestoft, ii. 259
— reaches home, ii. 260
— temporary rally, ii. 260
— his death, ii. 261
— Said Pasha's letter to him, ii. 261
— his funeral in Westminster Abbey, ii. 263
— permission given by the Queen to pass through Hyde Park, ii. 263
— expressions of regret, ii. 265, 266
— plate on his coffin lid, ii. 267
— legend on the monumental brass, ii. 267
— statue in St. Margaret's Gardens, ii. 267
— his report to the London and Birmingham Railway Company on the question of gauge, ii. 272

SUS

Stephenson, James, said to have been the discoverer of the blast-pipe, i. 146
Stephenson, Mr. George Robert, designs the Benha and Birket-el-Saba tubular bridges over the Nile, ii. 179, *note*
— his superintendence of the details of the Victoria viaduct, ii. 230
— superintends the iron work of the superstructure of the Victoria bridge over the St. Lawrence, ii. 209
Stephenson, Eleanor, i. 16
— her great disappointment, i. 16
— becomes housekeeper to her brother George, i. 16
— her 'artificials,' i. 17
— her marriage, i. 50
St. Lawrence river, the Victoria tubular bridge over the, ii. 180, 185
— breaking up of the ice on the, ii. 185, 186, 196
— detailed account of the Victoria viaduct, ii. 190
— Lake system of the waters flowing into the river, ii. 191
— winter phenomena of the St. Lawrence, ii. 196
St. Margaret's Gardens, Robert Stephenson's statue for, ii. 267
Stockman, Mr. B. P., his assistance in the designs for the railway bridges over the Nile, ii. 179, *note*
— sent to Canada to examine the progress of the Victoria bridge, ii. 226
Stockton and Darlington Railway, opening of, i. 111
— employment of locomotives on, i. 111
— Mr. Dixon's mode of fixing arches on the, ii. 289, *note*
Stokes, Professor, on the deflection of beams under moving loads, ii. 58
Suez, Isthmus of, surveyed by Robert Stephenson and M. Paulin Talabot, i. 240
Suez Canal, Robert Stephenson's speech on the proposal for, ii. 149
— discussion at the Royal Geographical Society, ii. 250
Sun-dial constructed by George Stephenson, i. 44
Surrey Iron Railway Company, incorporation of the, i. 265
Suspension bridges, ii. 37

SUS

Suspension bridges—*continued*
— Mr. Rendel's mode of trussing for preventing oscillation in the platform of, ii. 292
— Telford's bridge over the Menai Straits, ii. 77
Swannington, line from, to Leicester begun, i. 164
— coal in the neighbourhood of, i. 164
Swansea, George Stephenson's engine for drawing coals at, i. 70
Swinbourne, Mr., in Egypt, ii. 179, *note*
Switzerland, Robert Stephenson's account of railway progress in, ii. 184

TALABOT, M. Paulin, his intimacy with Robert Stephenson, i. 240
— their survey of the Isthmus of Suez, i. 240
Tamar, Mr. Brunel's iron bridge over the, ii. 67
Tate, John, the early companion of Robert Stephenson, i. 36
Tate, Robert, Robert Stephenson's visit to, ii. 240
Telegraph, electric, the commission ' To Inquire into the Construction of Submarine Telegraph Cables,' ii. 250
Telford, Thomas, his iron bridge over the Severn, ii. 34
— and over the Thames at Staines, ii. 35
— his experiments on the strength of iron suspended chains, ii. 39
— his bridge over the Menai Straits, ii. 39
— his ' Practical Essay on the Strength of Cast Iron,' ii. 44
— his formation of the great Holyhead Trunk Road, ii. 74
— his survey of the road from Shrewsbury and Chester to Holyhead, and of the Menai Straits, ii. 75
— appointed engineer of the Holyhead roads, ii. 76
— his suspension bridge over the Menai Straits, ii. 77
— his proposal for a new bridge at Newcastle-on-Tyne, ii. 114
Tequindama Falls, i. 109
Thames Tunnel in 1828, i. 115

VAU

Thompson, Thomas, of Black Callerton, i. 3
— gives George Stephenson his marriage breakfast, i. 5
Throwing the hammer, i. 42
' Times ' printing office, Robert Stephenson's visit to the, i. 71
' Titania ' yacht, the, destroyed by fire, ii. 170
— the *new* ' Titania,' ii. 170
— life on board the, ii. 241
Trail, Dr., the mineralogist, i. 75
Tramways, mode of laying down colliery tramways in the North of England, i. 216
— in 1676, ii. 3
Tramways in some of the mineral districts, ii. 4
Trent Valley Railway, expense of obtaining permission from Parliament to make the, i. 284
— iron bridges of the, ii. 92
— opening of the, ii. 106
Trent, iron triangular framed girder bridge over the, ii. 103
Trevithick, Mr., invites John Steele to join him, i. 24
— the memorable wager won by his engine, i. 24
— Mr. Wood's description of this engine quoted, i. 25
— encountered by Robert Stephenson at Carthagena, i. 105
— his peculiarities, i. 105
— proceeds with Robert Stephenson to New York, i. 105
Tubular constructions, details of the tubes of the Victoria bridge over the river St. Lawrence, ii. 219. *See* Bridges, iron
Tunnels, story of the gentleman who had conceived a passion for, i. 128, *note*
Tyne, the first ballast engine erected on the banks of the, i. 6

UNION suspension bridge over the Tweed, construction of the, ii. 38

VACUUM engine, Mr. Brown's, i. 34
Vallance, Mr. John, his experiments with atmospheric tubes, i. 301
Vaughan, Mr., in Egypt, ii. 179, *note*

VAU

Vauxhall iron bridge, ii. 36
Victoria bridge over the Wear, Robert Stephenson's, i. 258
Victoria (St. Lawrence) bridge, ii. 180
— Mr. Stephenson's speech on the bridge, ii. 184
— detailed account of this bridge, ii. 189, 216
— authorities consulted, ii. 189, note
— point at which it crosses the river St. Lawrence, ii. 189, 216
— engineering problems to be solved in its erection, ii. 195
— former surveys, ii. 198
— Mr. Ross's surveys, ii. 199
— report of Robert Stephenson and Mr. Ross, ii. 201
— approval of the plans by the assistant commissioner of Public Works, ii. 203
— visit of Robert Stephenson to the site of the bridge, ii. 203
— his letter to the directors on the subject, ii. 204
— the contracts for the bridge executed, ii. 209
— superintendence of the iron work of the superstructure, ii. 209
— site of the bridge, ii. 215
— width of the river at this point, ii. 216
— level of the river at different parts of the year, ii. 216
— bed of the river, ii. 216
— the approaches and piers, ii. 217
— height of the tube, ii. 217
— the ice-breaking planes, ii. 218
— the tubes, ii. 219
— sketch of the principal operations, ii. 220
— opening of the bridge to the public, ii. 227
— difficulties of the work, ii. 227
Victoria, Queen, opens the grand central railway station at Newcastle-upon-Tyne, ii. 139
— her expression of sympathy on the death of Robert Stephenson, ii. 263
Victoria sewer, discussion in the House of Commons on the, ii. 146
Vignoles, Mr., his support of the principle of atmospheric propulsion, i. 332
— his evidence on the Dee bridge accident, ii. 52

WES

Wales, Albert Prince of, his inauguration of the Victoria bridge over the River St. Lawrence, ii. 227
Walker and Rastrick, Messrs., their reports on the advantages and disadvantages of the locomotive system, i. 122
— recommend the adoption of stationary engines, i. 123
— their report on stationary engines as compared with locomotives, i. 150
— Robert Stephenson and Joseph Locke's answer, i. 151
— their report on the comparative merits of the locomotive and stationary engine system, i. 295
Walker, Mr. James, lays down a tramway in the Commercial Road, i. 227
— his examination before the Committee of the House of Commons, i. 227, note
— his evidence on the Dee bridge accident, ii. 85
Walker, on the Tyne, ii. 237
Walmsley, Sir Joshua, i. 164
Warrington and Newton Railway, i. 152, 154
Waterloo, Robert Stephenson's trip to, i. 114
Wear, iron arch bridge over the, near Sunderland, ii. 34
— Tom Paine's bridge over the, ii. 233
— description of it, ii. 234
— Robert Stephenson's restoration of it, ii. 70, 71
Wearmouth ferry, iron bridge at, ii. 34
Weatherburn, Henry, the enginedriver of the 'Harvey Combe,' ii. 266
Wellington, Arthur Duke of, at the opening of the Liverpool and Manchester Railway, i. 157
— his reception by the Lancashire mob, i. 161
Wellington Memorial School, in Newcastle, Robert Stephenson's donation to the, ii. 235
Wenham Lake ice, ii. 258
West Moor colliery, George Stephenson at, i. 10, 12
Westminster Abbey, Robert Stephenson's funeral in, ii. 263

WHA

Wharncliffe, Lord, his kindness to Robert Stephenson, i. 178
— presides at a meeting for the promotion of the London and Birmingham Railway, i. 179
Wheatstone, Professor, his friendship for Robert Stephenson, i. 235
— his electric telegraph, i. 235
— proposes Robert Stephenson as a member of the Royal Society, ii. 231
— at the Royal Society Club, ii. 252
Whinstone in the Snibston collieries, i. 164
Wigham, Anthony, his farm, i. 14
— his intimacy with George Stephenson, i. 26
— his farm at Long Benton, i. 26
Wild, Mr. Charles Heard, his evidence on iron bridges, ii. 58
Willington Quay, Hawthorn's ballast engine at, i. 6
— George Stephenson at, i. 6
— the Stephenson Memorial at, i. 10
Willis, Professor, his paper on the deflection of beams under moving loads, ii. 58
Willow Bridge colliery, George Stephenson embarks in the, i. 51

YOU

Wolsingham, Robert Stephenson's visits to, i. 18
Wolverton embankment, accidents at the, i. 195
Wood, Mr. Nicholas, his account of George Stephenson's experiment with his safety-lamp, i. 38
— Robert Stephenson apprenticed to him, i. 46
Woods, Mr. Edward, his particulars respecting the atmospheric system, i. 340, 341
Wye, Mr. Brunel's iron bridge over the, ii. 67
Wylam Railway, the memorable experiments on the, i. 31
Wylam, Mr. Hedley's mode of letting off the waste steam in his engines, i. 144
— Robert Stephenson's last visit to, ii. 240

YACHTS, Robert Stephenson's, ii. 170, 171
York and North Midland Railway, George Stephenson appointed engineer of the, i. 239
Young, the Hon. John, and the Victoria bridge over the St. Lawrence, ii. 180
— his suggestion in 1846 for a bridge over the river, ii. 198

THE END

LONDON
PRINTED BY SPOTTISWOODE AND CO.
NEW-STREET SQUARE

Printed in the United States
By Bookmasters